Robot Safety

International Trends in Manufacturing Technology

ROBOT SAFETY

Edited by
Professor Maurice C. Bonney
and
Dr. Y.F. Yong

IFS (Publications) Ltd

Springer-Verlag
Berlin Heidelberg New York Tokyo
1985

British Library Cataloguing Publication Data

Robot safety.—(International trends in manufacturing technology)
    1. Robots, Industrial—Safety measures
    2. Robots, Industrial—Design and construction
    I. Bonney, M.C.   II. Yong, Y.F.   III. Series
    629.8′92   TS191.8

ISBN 0-903608-69-3 IFS (Publications) Ltd
ISBN 3-540-15484-1 Springer-Verlag Berlin Heidelberg New York Tokyo
ISBN 0-387-15484-1 Springer-Verlag New York Heidelberg Berlin Tokyo

©1985 **IFS (Publications) Ltd,** 35–39 High Street, Kempston,
      Bedford MK42 7BT, UK
  and **Springer-Verlag** Berlin Heidelberg New York Tokyo

Phototypeset by Gilbert Composing Services, Leighton Buzzard, Bedfordshire
Printed and bound by Butler and Tanner Ltd, Frome, Somerset

International Trends in Manufacturing Technology

The advent of microprocessor controls and robotics is rapidly changing the face of manufacturing throughout the world. Large and small companies alike are adopting these new methods to improve the efficiency of their operations. Researchers are constantly probing to provide even more advanced technologies suitable for application to manufacturing. In response to these advances IFS (Publications) Ltd, is publishing a series of books on topics that highlight the developments taking place in manufacturing technology. The series aims to be informative and educational.

Subjects to be covered in the series include:

Robot vision
Programmable assembly
Robotic assembly
Robot safety
Robot sensors
Electronics assembly
Flexible manufacturing systems
Automated guided vehicles
Robotic welding

The series is intended for manufacturing managers, production engineers and those working on research into advanced manufacturing methods. Each book will be published in hard cover and will be edited by a specialist in the particular field.

This, the fourth in the series–Robot Safety–is under the editorship of Professor Maurice Bonney of Loughborough University of Technology, and Dr. Y.F. Yong of BYG Systems Ltd, Nottingham. The series editors are: Michael Innes, John Mortimer, Brian Rooks, Jack Hollingum and Anna Kochan.

Finally, I express my gratitude to the authors whose works appear in this publication.

John Mortimer,
Managing Director,
IFS (Publications) Ltd.

Acknowledgements

IFS (Publications) Ltd wishes to express its acknowledgement and appreciation to the following publishers/organisations for supplying some of the articles reprinted within this book.

British Robot Association
28–30 High Street
Kempston
Bedford MK42 7BT
England

Health and Safety Executive
McLaren Building
2 Masshouse Circus
Queensway
Birmingham B4 7NP
England

Arbetarskyddsstyrelsen
National Board of Occupational
  Safety and Health
Ekelundsvägen 16
S-171 84 Solna
Sweden

Robotic Industries Association
P.O. Box 1366
Dearborn, MI 481211
USA

International Flexible Automation
  Center (INFAC)
210 Century Building
36 S. Pennsylvania Street
Indianapolis, IN 46204
USA

Fraunhofer-Institut für Produktions
  technik und Automatisierung (IPA)
Postfach 800 469
Nobelstrasse 12
D-7000 Stuttgart 80
West Germany

Research Institute of Industrial
  Safety
Ministry of Labour
5-35-1 Shiba
Minato-ku
Tokyo
Japan

Machine Tool Industry
  Research Association
Hulley Road
Macclesfield
Cheshire SK10 2NE
England

IBM Corporation
P.O. Box 1328
Boca Raton, FL 33432
USA

Rolls-Royce Ltd
P.O. Box 31
Derby DE2 8BJ
England

Erwin Sick Optic-Electronic Ltd
Lyon Way—Hatfield Road
St. Albans
Hertfordshire AL4 0LG
England

U.S. Department of Commerce
National Bureau of Standards
Building 220
Washington, DC 20234
USA

Ford Motor Company
Central Office
Eagle Way
Brentwood
Essex CM13 3BW
England

Manufacturing Safety Coordination
Ford of Europe
Dagenham Training Centre
Dagenham
Essex RM9 6SA
England

Society of Manufacturing Engineers
One SME Drive
P.O. Box 930
Dearborn, MI 48128
USA

J.P. Udal Co.
Union Mill Street
Horsley Field
Wolverhampton WV1 3ED
England

Herga Electric Ltd
Northern Way
Bury St. Edmunds
Suffolk IP32 6NN
England

IFS (Conferences) Ltd
35–39 High Street
Kempston
Bedford MK42 7BT
England

IVF, The Swedish Institute
    of Production Engineering Research
Mölndalsvägen 85
S-41285 Gothenburg
Sweden

In addition, the Editors wish to acknowledge the help given by
Colin Thompson during the organisation of the two robot safety seminars held
by the University of Nottingham, in conjunction with the Ford Motor
Company, and to Mike Innes of IFS (Publications) Ltd for his thoughtful
editorial input and production work.

Last, but not least, grateful appreciation is due for the support given by
Claire Bonney and Zeynep Guvener during the production of this book.

I.F.S. (Publications) Ltd.

An International Fluidics Services Ltd. Company

35-39 High Street, Kempston, Bedford MK42 7BT, England

Tel: Bedford (0234) 853605   Telex: 825489

Our ref: Your ref: Date:

Dear Contributor

ROBOT SAFETY

Please find enclosed a complimentary copy of the book in recognition for
your most valued contribution. For your information, the book was published
on May 10 and co

Robot S
Manuf

Contents

1. Legislation and Standards

2. Surveys and Analyses

3. System Design, Implementation and Methodology

4. System Components

5. Case Studies

Preface

The objective of the book is to provide a worldwide and multidisciplinary framework to *robot safety*. The origins of the book were seminars on robot safety organised by the authors for the University of Nottingham, in conjunction with the Ford Motor Company. However, in the editing process we have set ourselves the more ambitious target of bringing together a state-of-the-art snapshot of robot safety from around the world and, to this end, we have included edited papers from the Federal German Republic, Italy, Japan, Sweden and the USA, as well as the UK.

This introductory overview reflects the Editors' personal perspectives. It sets the work of the various contributors in context and provides continuity between the diverse threads, legal, technical and human, which are relevant to this exciting field. The emphasis of the book is practical.

Scene setting: The need to plan robot safety
Robots are used increasingly to perform industrial tasks. In many cases robots form part of more complex systems, the components of which include not only the robot but other manufacturing machinery, conveyors, AGVs, controllers, sensors, and MAN. In particular, man is involved at many stages. He is planned to be involved in installation, programming and maintenance. He may also be in proximity as a worker at a nearby workstation, as a passerby or as a visitor being shown the new facility. It is when man and the robot system come together, in a planned or unplanned way, that potential safety problems arise. It is therefore important that the inter-relationships between man and the other system components are carefully designed. It is also important that hazardous, unplanned situations are avoided.

Robots differ from traditional machinery in that they have many degrees of freedom and are capable of a much wider range of movements. Robots use both hardware and software, together they form a complex system, faults in which can cause effects which are difficult to predict and are potentially hazardous. Accepting the need to plan robot workplaces carefully places the emphasis on the design of complete safety systems. This in turn requires understanding the system possibilities and the provision of appropriate training, and with such complex systems it is helpful to have procedures, rules, checklists, etc. to guide the design and operation of robot workplaces so as to minimise the chance of danger.

As is described later, many changes are occurring in the use of robots. The number of robots is increasing, the range of applications is increasing, the technology of the robot is changing, and there are some robots with increases in size, power and speed. Systems are also becoming more complex.

The relative newness and rapid rate of change of the technology means that legislation governing robot safety is general in nature, that codes of practice are recent and/or still being formulated, and that knowledge is dispersed. Thus,

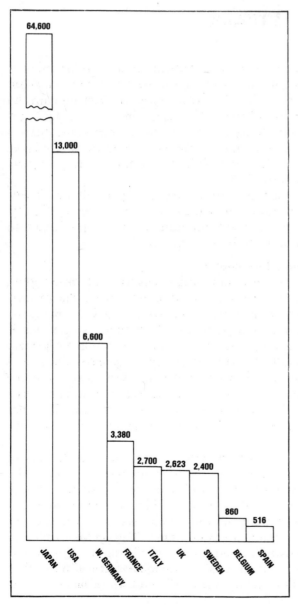

Fig. 1   World robot population

although great experience in the design of robot workplaces exists, it is widely scattered over many countries, different organisations and many different functional groups such as safety engineers, product and process designers, software engineers and consultants. Relevant journal articles and conference papers are equally dispersed. There are no standards as such, but there are general principles and there are tools at our disposal. This book brings many of the factors, legal, technical and human, affecting robot safety together in a structured way. In addition to the practical applications described within many of the papers a range of case studies is included to illustrate how other organisations are solving robot safety problems. The objective is to provide a state-of-the-art snapshot of this field of rapidly growing importance and to set these factors in a practical perspective.

Industrial robots and their applications

This brief section is summarised from the British Robot Association report, *Robot Facts 1984*. It quantifies the continued rapid growth in numbers and applications of industrial robots. Fig. 1 shows the distribution of robots worldwide, Fig. 2 the distribution between Japan, Europe and the USA, Fig. 3 the growth rates for the UK and West German robot population, and Fig. 4 the UK robot applications analysis. Of particular relevance is the diversity of the applications. It will be seen that a range of these will raise interesting protection problems. Energy sources other than the robot, such as furnaces and welding, raise one set of problems. Materials handling raises another set, and the potentially less structured situation arising in education and research raise yet other safety problems.

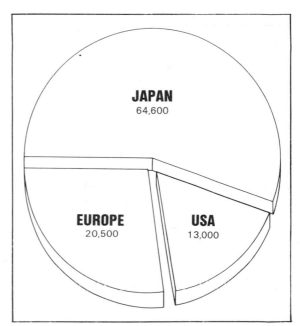

Fig. 2   Robot distribution between Japan, Europe and the USA

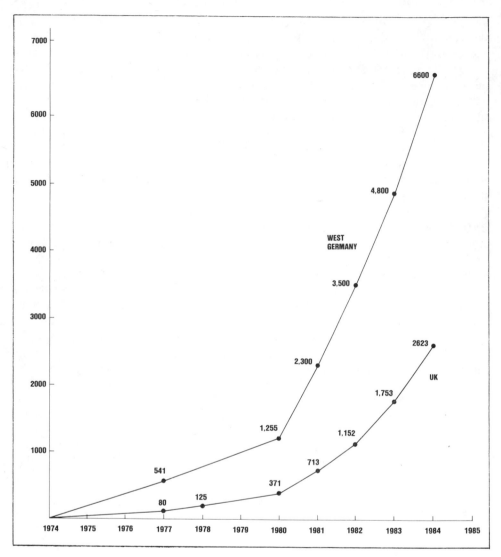

Fig. 3   UK and West German robot population

The sheer diversity of situations which can arise is one reason why we are unlikely to have cut and dried answers to robot safety and why there is the need to bring the best of current practice to bear on the problems.

The book organisation
The book brings together contributions on many of the factors which need to be considered when designing safe robot workplaces. The contributions relate mainly though not entirely to current practice. The technology, the applications and the codes of practice will each continue to develop as experience is gained.

	Installed during 1984	Total at Dec. 1984	% Increase during 1984
Surface Coating	24	177	15.6
Spot Welding	122	471	35.0
Arc Welding	107	341	45.7
Grinding/Deburring	16	43	59.3
Assembly	96	199	93.2
Die Casting	2	40	5.3
Injection Moulding	136	412	49.3
Machine Loading	48	213	29.3
Press Loading	11	59	22.9
Inspection/Test	11	41	36.7
Handling/Palletising	36	102	54.5
Education/Research	41	122	50.6
Investment Casting	0	14	–
Forging	3	10	42.9
Glueing/Sealing	9	23	64.3
Laser Cutting	1	2	–
Water Jet Cutting	3	3	–
Other Applications	13	27	–
Not Specified	0	133	–
Totals	679	2432	38.7

Fig. 4   UK robot application analysis

The book is divided into five sections:

- Legislation and Standards.
- Surveys and Analyses.
- Systems Design and Implementation.
- System Components.
- Case Studies.

Legislation and Standards. The context in which robot engineers, safety engineers, occupational health staff, etc. have to design their workplaces is described. In effect, the section outlines the constraints set by legislation, codes of practice and safety standards. It is clear that although there are some differences there is much in common between the different nations using robots.

Surveys and Analyses. Relevant work from West Germany, Sweden, UK and Japan is included. The papers provide some useful statistics of the nature and causes of industrial robot accidents.

System Design, Implementation and Methodology. This section examines the design of safe robot systems in very broad, top down terms. It looks at safety systems for the future, at real-time risk assessment by a separate safety computer, at designs for safeguarding, at CAD as a robot workplace design tool, at a safety watchdog computer which protects the robots, at the hardware and software of the robots themselves, and how these are designed with safety in mind. Finally, it examines training requirements for the safe and effective implementation and operation of industrial robots.

System Components. This section deals with the individual items which may be used as part of the robot safety system. These include perimeter guarding, safety interlocking, photoelectric guarding, safety mats, safety sensor systems and a robot safety and collision avoidance controller.

Case Studies. A collection of case studies describing robot installations is presented. Two of the case studies are by manufacturers of safety equipment, the other three describe users installations. Together with the applications described in earlier sections they highlight the problems of robot safety.

A look into the future

The general framework of the book will now be clear. Between the various sections the papers provide a thorough 'state-of-the-art' overview of many of the factors which affect robot safety. Yet there are some important topics which have not been given the coverage or weight which they deserve. There are also many technical developments which mean that while new problems will arise, the design possibilities for safe robot workplaces will continue to expand and offer scope for novel solutions.

Some important developments include:

- Robot numbers are increasing.
- Robot applications are extending.
- Robots are acquiring enhanced hardware facilities (e.g. increased power, speed, precision, sensors, and more powerful computer control).
- Robots are acquiring enhanced software facilities (more powerful computers allow more complex programs, editing, interrupts, etc.).
- Robots are becoming part of larger systems often computer controlled (e.g. part of an FMS cell and part of a networked automatic production facility).

Taking each of the first four sections of the book in turn we see that:

- Legislation and standards are changing.
- Statistics are increasingly available.
- System design methods are developing.
- The system components are changing and becoming more reliable.

In particular, our knowledge and understanding of robot systems and methods and involvement in training are improving.

It is clear, therefore, that things are changing and, although the exact details and their timing may be conjectural, the broad brush of development is discernible. Statistics will improve and this alone will influence actions that are taken in the future. There is the possibility that robots, especially if part of a computer controlled cell, can be self-monitoring. In addition to this evolutionary development there are considerable changes to be expected under the broad heading of computer integrated manufacture. First the software in the robots and in any supervisory and safety computers will be enhanced and generally made more reliable. Downloading of software from a supervisory computer to the robot will be routine and off-line programming will become regularly used. Together these developments will make robots more suitable for small batch manufacture. Thus we see a cycle of influences. Off-line programming will not only reduce the safety risks associated with programming but improve productivity by not taking the robot out of action when programming is performed. This allows the robot to be used economically for small batch manufacture. This together with the downloading of programs from some central database and the use of standard procedures for automatic tool changing means that the way is opened for robots to operate routinely within a truly flexible manufacturing system.

Note, though, that if the above occurs the safety problems are changed and we now need supervisory computers and the use of rules for scheduling and controlling the manufacturing system. Although the system is more complex there are fewer people, who will be used mainly for maintenance – one of the risky operations. Maintenance should preferably occur while the rest of the system is still in operation and would require careful safety planning.

The use of sensors will enhance the capabilities of the robot allowing more adaptable, less deterministic, systems to be created. But they will also allow the presence of people within the robot's hazard zone to be identified and this, in turn, perhaps after appropriate risk assessment, could bring appropriate safety procedures automatically into operation. However, equipment will still break down and maintenance will still be required and restart conditions must be appropriately coordinated.

The potential danger of using a near automatic system of great complexity is that the systems will be very difficult to comprehend and so their introduction will need to go 'hand in glove' with cautious safety strategies if they are not to cause human safety problems despite their relatively small use of people. Conversely, the developments outlined above offer tremendous opportunities. We do not want these cautious safety strategies to unnecessarily inhibit developments.

Back into perspective

The major factors are hardware, software, people and costs. With the technology we can make robot systems as safe as we wish provided there is sufficient investment in design, training, implementation and operation. The presence of people means that the man–machine interface needs to be ergonomically designed. This has implications from improved teach pendant design, to the development of 'simple' off-line programming systems, to the use of transparent diagnostics, to appropriate training and retraining, to suitable organisation structures right through to safety codes of practice and legislation, each consistent with safety *and* flexibility. Accidents occur because of incompatibility between the designed system and the capabilities of man. Ergonomics, although not specifically covered in the papers must not be neglected.

It is essential that as technology moves forward, the costs, the opportunities, the problems and the solutions will all change. The various sections of the book provide a good insight into the way things are moving indicating different solutions but common objectives and increasing knowledge and awareness so that within the practical constraints robot systems are made as safe as possible.

Although some important factors have not a separate paper which deals specifically with one particular aspect, e.g. the installation and the programming of robots, they are discussed in other papers in passing. All stages of the construction and operation of safe robot installations are in the interest of, and the responsibility of, the robot equipment manufacturers and the users.

Maurice C. Bonney
Y. F. Yong
April 1985

1
Legislation and Standards

This section describes the context in which robot engineers, safety engineers, occupational health staff, etc., have to design their workplaces. In effect, the section outlines the constraints set by legislation, codes of practice and safety standards. Although there are some differences, it will be seen that there is much in common between different nations.

The first paper is by James Barrett of the UK Health and Safety Executive. Here, he seeks to point out the principal legal requirements of which suppliers and users of robots should be aware when considering the safeguarding of installations. Following a brief description of some of the problems of robot safety he notes that the international nature of the advance of the robot industry has not to date produced accepted standards, although the interest in health and safety in the UK may also be found in the USA, France, Germany and Scandinavia. The legal requirements pertaining to the UK and robots are then outlined, and the complexity of the underlying problems is nicely illustrated in, for example, his comments on risk assessment. The paper ends with a checklist of questions which users of robots will find useful in considering their installation and safety requirements.

In the second paper, Neil Percival examines safety standards in a worldwide context and includes specific comments on design, safeguarding and user requirements. Percival shows not only the commonality between broad developments worldwide, but also that the detail within that picture is changing rapidly. Generally, standards or draft standards are emerging in many countries and there is considerable international cooperation. In order that the content was as up-to-date as possible, Neil Percival revised his contribution in March 1985 – the print deadline.

However, new documents are continuing to appear and the Editors would have liked to have included in this book several further items which are to be published shortly, e.g. the proposed American Safety Standard produced by the Robotics Industries Association, USA. Although the timing for this was a little premature it provides a useful reminder of the rate of progress.

Robot Safety and the Law

R.J. Barrett
Health and Safety Executive, UK

An outline of the major safety legislation likely to affect robot manufacturers, users and suppliers is given, and suggests that with a systematic assessment of the possible hazards the safe introduction into industry of this device can be achieved.

The introduction of robots into industry raises questions regarding both the means by which safety in their use may be assured and the application of existing legislation to those means. There is an evident need for the preparation of guidance and some attempts are being made to prepare this. In the meantime, however, the use of robots is increasing and is likely to continue to do so in spite of economic difficulties. Our most urgent need, therefore, is to have some sort of coordinated approach to the problems created.

Robots presently in use differ from other machinery and plant in that they have several degrees of free movement, and this raises some special problems which may vary according to the robot's use. Assuming proper attention is paid to all the safety considerations which are found generally in industry, additional factors may need to be considered in relation to a robot:

- The safety of persons working in the general vicinity.
- The presence of fixed objects (e.g. stanchions) within the robot's area of operation.
- The close proximity to the robot of programmers, diagnosticians, maintenance personnel, cleaners and others in the normal course of their duties.
- The robot's relationship to other machinery or plant.

State-of-the-art

The general approach to safety which has been applied by most known users of robots has been to fence them off in some sort of interlocked enclosure. Standards vary; where the intention is merely to exclude unauthorised persons, most users provide orthodox 2 m high fencing with interlocked gates, whilst others provide lower fencing as an indication that access is restricted. There may be circumstances where either of these approaches may be acceptable but neither will provide protection for the programmer, diagnostician or setter, who needs to enter the enclosure when power of some sort has to be applied to a robot. Where a robot is used in connection with other machinery (e.g. transfer mechanisms), an orthodox interlocked enclosure may be required because of the dangers from the other machinery, and the terms under which entry into the enclosure can be permitted for servicing the robot have to be decided on the facts of each case. In other circumstances (e.g. with a free standing robot) the robot may offer no significant hazard to persons other than those who have to work directly on it, for example in programming or fault finding.

The hazards to which a programmer is likely to be exposed may include aberrant behaviour by the robot. The possibility of this happening can be reduced by careful design although it is unlikely to be eliminated. The effects of aberrant behaviour can also be limited by the use of, for instance, stops or buffers to reduce the amount of free movement, or by using lower speeds of robot movement during programming or fault diagnosis. In the future we may look forward to robots which are intrinsically safe, but for the present the prospect of eliminating hazards altogether or of providing secure fencing to prevent injury to persons who service robots seems unlikely.

If physical safeguards are not practicable we have a dilemma. The free moving arm of a robot and its grippers may clearly be dangerous parts of the machinery and thus require safeguarding. Certain operations may, of necessity, need to be carried out while a person is in a position of danger and there may be no single viable means of ensuring safety for a person who has to work on a robot which he may have to manipulate under power in order to carry out his task. A combination of measures such as hold to run controls, presence sensing devices, physical restrictions of robot arc of movement and speed limitations can lead to a reduction of risks. In the longer term a solution could be found by applying a sensor on the moving part of the robot which will inhibit motion if a human being is in the path of the robot. No such device appears to be in use at present. The increased use of off-line programming may also provide for safer working but lead-through programming still seems to be the most popular method of teaching.

The industrial robot is subject to an existing framework of safety legislation which will be examined. The robot may be seen as presenting novel problems for those responsible for its safe introduction but clearly it has a potential for safety benefits as well.

This paper seeks to point out the principal legal requirements of which suppliers and users of robots should be aware when considering the safeguarding of installations. An attempt has been made to include a number of suggestions which will, it is hoped, assist users in assessing the requirements of safeguards and the level of integrity which should be applied to the control system to ensure the safety of operation personnel.

It may seem strange, at first, that safety is causing so much interest when robots so clearly have a potential for removing people from hazardous and unpleasant areas and relieving them of arduous tasks. One possible reason why the subject is causing so much excitement is not so much related to its capacity for harm but to the fact that there has been a recognition within society that new developments are often accompanied by problems which were not considered at an early stage of design. One has only to consider the early factory health and safety legislation to see that this was developed as a response to conditions. Laws were passed to remedy poor conditions which were becoming apparent and not to prevent them happening in the first place. Society today does not allow us the benefit of hindsight and in any case the development of safety at a design stage is highly desirable from both a legal and an economic viewpoint. The advantages for the user, of course, are that not only will the plant be safer, and therefore less likely to lead to industrial problems, but this approach proves less costly than add-on safety requirements. The development of safe design whilst welcomed is unlikely in itself to remove all machinery dangers and so there is still a need for an assessment of the hazards to be made by both the manufacturer and the user.

At present, the UK robot population is still small by world standards but there are signs that the numbers are about to increase significantly. There will also be found new areas of usage and current developments with possible legal implications include mobile robots and some of the developments on sensors. It would be quite wrong to ignore the benefits which robots bring with them including the possibility of removing the need for operators to place their hands between the platens of presses and diecasting machines or work in toxic/flammable atmospheres. The HSE does not wish to inhibit the use of robots but there is a growing amount of evidence that accidents are occurring and experience has shown that if safety is given proper consideration at the early stages, not only are the costs reduced, as mentioned earlier, but the final safety solutions are more elegant and effective. Inspectors of Factories are always prepared to help by discussing the legal issues affecting installations, but the responsibility for ensuring safety remains with those who design, manufacture, import, supply and use them. This fact, plus the rate at which robots are being introduced and the range of applications, means that suppliers and users will have to make decisions on safeguards themselves and many will have little experience to draw upon in reaching their decisions. There will also be companies who for the first time will feel the effect of legislation in safety fields.

Traditionally, safeguarding standards have developed over a period of time. The international nature of the advance of the robot industry has not to date produced accepted standards although the interest shown in health and safety in this country may also be found in the USA, France, Germany and Scandinavia.

The supplier, manufacturer, user and installer of a robotic installation should be aware at the very least of the relevant legal requirements which apply and the standards of safeguarding which are considered acceptable for similar risks with existing machinery. These requirements are outlined but those looking for a simple specification for robot safeguarding will be disappointed. It is possible, however, to determine an appropriate level of safety in a systematic way.

Legal background

In the UK there are two major Acts of Parliament relating to health and safety; these are the Factories Act 1961 and the Health and Safety at Work Act 1974. In addition to these major items of legislation there are numerous Regulations and this can mean that for a particular robot application a large number of legal requirements may be relevant. For example, robot paint spraying may involve consideration of the Highly Flammable Liquids and Liquefied Petroleum Gases Regulations, while robot grinding may involve Abrasive Wheels Regulations, robot fettling could involve the Grinding of Metals Regulations, and all, if placed in factories, will be subject to the Electricity Regulations. It is not intended to discuss these Regulations in great detail except to say that where there are areas of safety legislation which appear to be particularly relevant they should be considered.

Of the two major Acts mentioned above a clear distinction may be drawn between the specific nature of the Factories Act (in effect a list of requirements) and the more general nature of the Health and Safety at Work Act, which is not only more widespread in terms of those it protects (the Act covers employed persons rather than factory premises) but is more widespread in terms of the allocation of duties and places the emphasis for the assessment of risks and for the implementation of appropriate safety measures on to manufacturers, suppliers and users on the basis of what is 'reasonably practicable'. This principle of extending the responsibility for determining what is safe and applying the necessary measures to ensure safety is a feature of more recent legislation. This does not mean, of course, that guidance on acceptable standards will not be available, but manufacturers and users alike now have broad responsibilities and the effectiveness of their response to these duties will be judged ultimately not by the Factory Inspectorate but by the Courts.

It is assumed that if dangers do exist with robots they will be of a safety nature rather than a health nature and so safety legislation alone will be considered.

Factories Act

The major sections of the Factories Act[1] which deal with safeguarding are Sections 12 to 16. Sections 12 and 13 deal with prime movers and transmission machinery and these normally present no problem. Section 14 deals with machinery safeguarding: sub-section (1) requires "every *dangerous part* of any machinery . . . shall be *securely fenced* unless it is in such a position or of such construction as to be as safe . . . as it would be if securely fenced". Sub-section (2) allows an automatic device to be used in lieu of a fixed guard if the nature of the operation precludes a fixed guard. The interpretation of sub-section (1) has been assisted by much case law and legal argument has centred on the test of: What is a dangerous part? What is secure fencing?

"A part of the machine is dangerous if it is a reasonably foreseeable cause of injury to anybody acting in a way in which a human being may be reasonably expected to act in circumstances which may reasonably be expected to occur".[2] So when we consider danger, we also have to consider what is reasonably foreseeable?

The standard of secure fencing requires the result to be achieved. Fencing is secure which effectively protects the workman from the danger of contact with the exposed parts of the machine. The section is rightly regarded as one of the most effective safety provisions and can be seen as providing an almost absolute duty once a dangerous part has been demonstrated. To illustrate this point consider the Woodworking Machine Regulations. It is necessary to approach a revolving part of machinery to cut wood. The part must be exposed and so Regulations were required which specified alternative safeguards to Section 14 (1). If the Woodworking Regulations were not in force these machines could not legally be used because they would not meet the requirements of Section 14 (1). The same principle applies to Abrasive Wheels.

Section 14 (1) does not require the fencing to protect against fragments, if shattered, flying out of a machine or ejected from a robot, and requires fencing of 'parts' of 'machinery' and not the machinery as such.

Section 15 deals with an exception for certain operations at parts of machinery normally considered safe by construction or position, providing certain conditions are met. These conditions are contained in the Unfenced Machinery Regulations 1938.

Section 16 gives requirements for the standard of fencing with an exception allowed for certain operations which required guards to be removed while the machinery is in motion or use. As in Section 15, this exception is granted for certain operations under conditions specified by the Unfenced Machinery Regulations. These operations are the examination or any lubrication or adjustment shown by the examination to be *immediately* necessary if the examination, lubrication or adjustment can only be carried out while the part of the machinery is in motion or use. One may consider whether any of the necessary approaches to a robot could be described as examination, lubrication or adjustment or whether the robot is in motion or use. An example could be the close observation required for welding robots. It is, however, important to establish immediate necessity rather than convenience.

Health and Safety at Work Act
More general requirements are contained in the Health and Safety at Work Act 1974[3]. Section 2 of this Act deals with the general duty of employers which is to ensure, so far as is reasonably practicable, the health, safety and welfare at work of employees. This duty extends to, amongst other things, providing and maintaining plant and systems of work which are, so far as is reasonably practicable, safe and without risks to health.

Section 6 places a duty on any person who designs, manufactures, imports or supplies any article for use at work to ensure, so far as is reasonably practicable, that the article is so designed and constructed as to be safe when properly used. To carry out this duty it may be necessary to arrange for examinations and testing of equipment and to make available adequate information for users.

The phrase 'so far as is reasonably practicable' appears quite often and the following extract from one of the HSE publications[4] may assist in understanding its meaning. "Someone who is required to do something, so far as is reasonably practicable, must assess on the one hand the risks of a particular work activity or environment and, on the other hand, the physical difficulties, time, trouble and expense which will be involved in taking steps to avoid the risks. If, for example, the risks to health and safety of a particular work process are very low, and the cost of technical difficulties of taking certain steps to avoid those risks are very high, it might not be reasonably practicable to take those steps. However, if the risks are very high, then less weight can be given to the cost of measures needed to avoid those risks. The comparison does not include the financial standing of the employer. A precaution which is reasonably practicable for a prosperous employer is equally reasonably practicable for the less well-off."

Robot applications are likely to fall within the scope of both of the above Acts which should not be seen as requiring two standards but covering all places of employment and work activities.

Safety framework

As stated earlier, whether or not robots present danger depends upon the facts of each particular installation. The size, positioning, method of programming and use will all affect an assessment of hazards. The question of danger, therefore, and of compliance with the law will be considered in respect of each installation by HM Inspectors of the Health and Safety Executive (HSE). Both HSE and industry would wish problems arising from the use of robots to be anticipated and acted on, through joint investigation by industry and HSE, and through discussions on the practical, technical and legal issues involved rather than their coming to light only as a result of serious accidents.

During the Brighton Conference, 'Automan 81', HSE introduced a safety framework[5] developed along the principles outlined in the British Standard Code of Practice BS 5304: 1975, *Safeguarding of Machinery*[6]. This standard identifies and describes methods of safeguarding which may be applied to dangerous parts of machinery and indicates the criteria to be observed in the design and construction and application of such safeguarding. If we consider that a robot presents a danger but that each installation needs to be assessed on its merits, there is a clear implication here that the methods of remedying any danger will vary along a scale of risk assessment. The suggestion made by the HSE paper was that the BS 5304 basis of risk assessment could be adopted as a yardstick on which to base further analysis of safety problems. BS 5304 introduces the ideas of fixed guards and interlocked guards and also considers electrical safety systems where the integrity of the safety system is also assessed.

The result is two standards of electrical interlocking, one appropriate to high-risk situations (high-risk interlocking) and one appropriate to other situations (normal interlocking). The question of whether to adopt a high-risk or normal-risk interlocking is determined in any particular cause by taking account of the relative importance of all the factors involved, and it is suggested that a similar assessment may be made in respect of robot installations where guarding is required.

The factors determining the risk involved include:

- The frequency with which access to the danger area is required.
- The foreseeable risk and severity of injury should the interlock fail, taking into account the method of working, the likely need for access, the action of the part safeguarded by the interlock, and the characteristics of the machine.

If interlocked fencing has been chosen as the guarding method the safety circuits should preferably operate directly on the electrical control or power circuit elements rather than via the internal logic of the system.

If we assume that a robot is working with another machine (or machines), we should consider the hazards brought about by the robot itself, the machine itself, and then the robot and the machine together. It is then necessary to consider the principle modes in which the robot will be working, and although these may vary, one could consider that programming, normal working and maintenance operations are the three main modes. In the case of extensive robot developments one could also consider installation. For each of the principal modes of operation one should then make an assessment of both 'designed' and 'aberrant' behaviour.

The term 'designed' behaviour is self-explanatory and means that the system is operating in its intended manner. The term 'aberrant' behaviour refers to any unpredicted movement of the machine system caused by a malfunction of the control system. A robot may, for example, behave aberrantly as a result of electrical interference or because an error may have been introduced into the program. This error may have been inherent in the program as conceived or may have been introduced as a result of some transient fault. The control system may have been affected by a severe environment.

Clearly, control systems of modern machines have high degrees of reliability. The test, therefore, is related to the reasonable foreseeability of aberrant behaviour, the consequences of such behaviour and what is reasonably practicable. The introduction of computer control[7] to robotics has enhanced flexibility but may also have introduced a potential for malfunction in, possibly, an unpredictable manner. This may have no safety impact at all but where it does then a number of options may be available including an improvement in the control integrity or assuming that the fault will occur, and minimising any consequent damage. Of course there may be differences of opinion on what are appropriate measures but the very fact that an assessment has been made is likely to enhance general confidence in the robot.

The nature of modern robots would appear to rule out close contour fencing, so the realistic options are enclosure type fencing, sensors, increased control integrity or simply to do nothing. It is obviously not sensible to remove existing traps and then create others between the robot and fencing which has been erected.

A great deal of the potential problems with robots will be removed by developments in the way in which robots are programmed and the sensors which are applied to them. One of the most encouraging aspects of robot development from the safety point of view is the interest shown in the subject by both users and suppliers and the broadly similar attitudes shown towards safety so far by France, Germany and the UK. As a demonstration of the interest of HSE in the topic it has become involved in European talks aimed at cooperating on robot safety and recently agreed to help form a robot safety liaison group to coordinate UK thinking on robot safety. This group will not be concerned solely with the European talks but will be in a position to advise on the safety aspects in this important area of industrial development.

The legislation on safety in this country is not normally directed at a particular target and it is not expected to see Regulations passed in the near future. The flexibility allowed by such an approach should be welcomed by industry as a means of developing safeguarding systems to take into account the legal, technical and practical problems which are faced.

HSE Checklist

A list of questions which users of robots may find helpful in considering their installation and the safety requirements is given in the Appendix to this paper This was developed within HSE as part of its contribution to the work of the MTIRA on a Robot Safety Code of Practice. Nevertheless, the questions will have a wider relevance.

The first section is concerned with the environment, for people to consider whether the atmospheres are flammable and whether the robot is required to be suitably explosion protected. The question of extremes of temperature may be relevant to the functioning of the control system. Corrosion, dust and fume, vibration and noise are all factors which may have a bearing on the safety of the system and where these factors are present the design should take these into account. The control equipment should be protected against electrical interference; this is fairly commonplace these days, but incidents are still reported to the HSE of conducted interference on the main supplies, the effects of electrical storms and electromagnetic fields, and so again one would expect adequate filtering and possibly other measures to ensure that the robot control system is not susceptible to these sorts of electrical effects. Although at an early stage, work has started, within the BSI, on recommended electrical interference immunity levels and this work should be of use in this area.

The second major area to consider is the location and how accessible the robot and its control system are to personnel who are not directly involved in its operation. Measures may be needed to ensure adequate security of the installation or stored data from casual, inadvertent or deliberate interference. If the robot is placed in a work area which contains static objects which could give rise to crushing or trapping points, these must be considered. Additional hazards may be introduced by the juxtaposition of the robot and associated machinery or plant, and again the robot installation may involve machinery for which recognised standards exist. If enclosure guards are used they should protect against the most likely form of accident. Ejection of material, weld spatter, and ultraviolet glare may need to be considered as well as more conventional trapping or entanglement risks. The presence of the robot may affect the safety interlocking of other equipment and this may be overlooked.

The third area which HSE believes should be looked at involves an assessment of which personnel need access to the robot. They may be skilled electrical or mechanical plant maintenance engineers and programmers, or service engineers, perhaps from the robot manufacturers, as well as quality control and factory cleaners. All of the tasks associated with the installation which require approach can then be identified and the category of person involved in these tasks chosen.

The installation may contain stored energy in the form of pneumatic reservoirs. The effect of loss of power or interruption of power on the system should be assessed.

The robot specification should then be studied. If the robot is operating within a factory and is pneumatically powered, any air receiver associated with the equipment will be subject to the requirements of Section 36 of the Factories Act. These requirements include periodic, thorough cleaning and examination and the fitting of a specified number of safety devices, e.g. pressure gauge, safety valve, drain plug and suitable means of allowing the interior to be thoroughly cleaned. Robots used for lifting are unlikely to be classified as cranes or lifting machines but there may be occasions when the use of the robot requires some consideration to be given to the safe working load.

Of course, one of the most important questions to ask is whether the size and the energy involved in the robot gives a reasonably foreseeable risk of injury occurring. Speed of movement and stopping times may also be relevant. On the programming side, there will have been much development work done by the manufacturer including program checking systems and there may be data available on electrical interference susceptibility tests. It may be desirable to have an indication as to whether the robot is at the end of a workcycle or still within a program. There may be quite lengthy pauses to allow for processes to take place which could lead observers to think that the robot was in a safe condition.

Another area for consideration concerns emergency stops. There will be times when a complete power shut-down is required at an installation and this may in some cases lead to loss of positional data. In other cases a hold position may be preferred, but with a facility for a complete isolation of power for maintenance purposes.

When considering programming by the user, proper documentation and security of programming could be important. Where the programming involves the use of a mobile teach box the control layout should be designed in such a way as to ensure the controls are clearly marked and well spaced to avoid inadvertent operation. It would seem desirable for the use of a teach box to remove any alternative control from the robot automatically. Teach restrictions and deadman's handle-type controls would also seem to enhance the safety of the programming operation and are to be recommended together with low torque, speed, etc.

It would be quite unrealistic to expect to find tailor-made solutions for any guarding problems which may be associated with the development of robotics. There is, however, a willingness on the part of HSE to discuss fully with all concerned in the industry the potential problems with a view to achieving practical solutions. Developments in robotics can make a contribution to safety if the approach taken by those concerned with this industry remains responsible. The important questions which users should ask are not "Do we have to guard our robot?" but rather "Is there a reasonably foreseeable risk of injury associated with aspects of robot use?" Answering that question will likely lead to the safe and responsible development of this important industrial tool.

References

1. *Factories Act,* 1961. HMSO, London.
2. Lord Guest, 1962. *Close vs Steel Company of Wales* (AC 367).
3. *Health and Safety at Work Act,* 1974. HMSO, London.
4. *A Guide to the HSW Act,* HS(R)6. Health and Safety Executive, Birmingham, UK.
5. Barrett, R. J., Bell, R. and Hodson, P. H. 1981. Planning for robot installation and maintenance: A safety framework. In, *Proc. 4th British Robot Association Annual Conference,* pp. 13–28. British Robot Association, Bedford, UK.
6. BS 5304: 1975. *Code of Practice: Safeguarding of Machinery.* British Standards Institution, London. (Currently under revision).
7. *Microprocessors in Industry–Safety Implications of the Uses of Programmable Electronic Systems in Factories.* Occasional Paper Series OP2. HMSO, London.

Appendix–HSE Checklist

Environment
- What environment is the system to be placed in?
  - –Is the atmosphere flammable? Is the robot required to be explosion protected, e.g. intrinsically safe, pressurised? (see BS 5345)
  - –Is the robot expected to operate extremes of temperature?
  - –Is the atmosphere corrosive?
  - –Are there problems of dust and fume?
  - –Is vibration, noise, etc, present?
- Has the design of the system taken into account any of the above factors which may be relevant?
- Is the control equipment protected against electrical interference?
  - –Is there conducted interference on the main supplies?
  - –What are the effects of electrical storms?
  - –What are the effects of electromagnetic fields and electrostatic effects?
  - –Is adequate filtering fitted to the power supplies?
  - –Is the environment electrically 'noisy'?

Location
- How accessible is the robot and its control system to personnel not directly involved in its operation?
- What measures are needed to ensure adequate security of the installation or of stored data from casual/inadvertent/deliberate interference?
- Does the work area contain static objects which could give rise to crushing/trapping points, e.g. stanchions?

- What, if any, additional hazards are introduced by the juxtaposition of the robot, and associated machinery or plant?
- What effect does the robot installation have on recognised machinery safety standards? (e.g. the Zinc Alloy Diecasting Association standard for pressure diecasting machines; the British Plastics Federation standard for injection moulding machines; MTIRA standards on transfer machinery in the motor industry).
- Does proposed guarding prevent likely forms of accident? (e.g. ejection of material, weld spatter, uv glare, as well as trapping).
- Does the presence of the robot affect safety interlocking of associated machinery? (e.g. transfer mechanisms).

Hazard assessment
- Which personnel need access to the robot?
  –Electrical/mechanical plant maintenance engineers?
  –Programmers?
  –Service engineers?
  –Quality control?
  –Cleaners?
  –Others (specify)?
- Have all the tasks associated with the installation which require approach been identified? Has the category of personnel involved in these tasks been chosen? Has any limitation been placed on their work?
- Does the installation contain stored energy (e.g. pneumatic reservoirs)?
- Has a list of potential hazards been drawn up? Does this include possible malfunctions?
- What is the effect of losses of power or interruption of power on the system?

Robot specification
- Is the robot operated hydraulically, electrically, pneumatically?
- What is the speed of movement of the robot?
- Is the safe working load of the robot arm significant and is this clearly marked?
- Taking into account the size and energy of the robot and its foreseeable actions is there a foreseeable likelihood of injury occurring?
- If yes, can a reduction of energy or alterations of position of the robot substantially reduce hazard?
- Can the robot energy be reduced during the times when personnel are approaching the robot?
- Can movement be restricted/minimised/halted by physical restraints at the extremes of the required robot movement?
- If the speed of movement is restricted, is this achieved by: control circuitry; primary power source reduction?
- Can the presence of a human be detected by the robot through sensor mechanisms?
- Do grippers retain the tool/workpiece during emergency stop or power loss situations?
- Can 'hold in the last state' control signals be over-ridden if, for example, a gripper has trapped an operator's arm and needs to be operated manually to release it?
- Has an assessment been made of the consequences of failure of the robot control elements (e.g. valves, motors, etc.)

Programming by manufacturer

Considering the basic robot program supplied by manufacturer:

- Is the main memory held in a non-volatile form?
- Are specific safety interlocks contained in software?
- If so, why was the above decision taken?
- Are program checking systems used during the testing of the robot and during normal running?
- Does the robot indicate to the observer whether it is at the end of a cycle or is still in the programmed mode?
- What diagnostic facilities are available?
- What assessment has been made of the effects of loss of data during transfer of information?
- Can the robot be moved from any point in the sequence to any other programmed point with a possible undefined pathway?

Programming by the user

- In what form is the program held?
- Is more than one program held in a memory at one time?
- What is the security between programs?
- Is access to program changing easy and what is the possibility of corrupting the existing data?
- Is the machine taught under power?
- What documentation exists for recording programs and amendments?
- How many people have responsibility/authority to change or load programs?
- What training is given?
- What safety assessment is made of programming procedures?
- Does the programming involve the use of a mobile teach box? If so, is the control layout ergonomically designed and are the controls clearly marked and well spaced to avoid inadvertent operation?
- Does the use of a teach box automatically remove control from the main console?
- Can a complete cycle be initiated from the mobile control box?
- Is the power control of a deadman's handle type?
- Does a teach power-restrict device operate?
- Does the restriction of power act through the control circuitry or directly affect the power?

Aberrant behaviour

- Has an assessment been made of the possible consequences of control malfunctions, e.g. unprogrammed movement, release of jaws, change of speed?
- Is the memory system protected in the event of power loss and can any safety backup system be exercised and checked?
- Does the robot manipulate hazardous materials (e.g. explosives, acids, hot articles, molten metal, radioactive materials) and are the backup/emergency systems appropriate for the additional degree of risk?

General considerations

- Is close observation necessary during a powered run of the robot?
- Are dry runs carried out?

- Has safeguarding been considered in relation to those persons who may need to work in the vicinity of the robot? For example,

Low power (speed)/sensor systems	–programmers
Interlocking	–maintenance engineers
Fixed fencing	–passers-by
Pressure mats	–unauthorised entry

These examples above are not meant to indicate which system is approved or appropriate.
- What interfacing is carried out with other machinery? Are the safety interlocks processed by the internal logic?
- What monitoring is carried out by the system, e.g. overspeed, overtravel, overheating, time to execute checks, positional feedback?

Safety Standards in Robotics

N. Percival
Machine Tool Industry Research Association, UK

The state-of-the-art on the development of safety standards for robots in the UK and overseas is described. It is clear that the subject of safety standardisation for robots is in a state of flux which is understandable for any new technology. However, it is also clear that the UK has established a useful reputation in this field and is in an excellent position to influence developments both within Europe and worldwide.

With any new technology it is necessary to pinpoint the hazards involved in using that technology, to assess the risks and then to apply established and proven methods of safeguarding to minimise the hazards. In addition, consideration has to be taken of the legal responsibilities of the designer, manufacturer, supplier and user of the equipment.

The UK has taken the lead in publishing industry standards through the guidance of the Machine Tool Trades Association (MTTA) and the safety framework of the Health and Safety Executive (HSE), both of which are having considerable impact overseas and are used as the basis of similar foreign standards. West Germany has published a VDI recommendation, Japan a general code of safety, and France and the USA are producing similar standards.

UK activities

The recent MTTA document[1] prepared with HSE assistance is not a standard in the accepted sense of, say, a British Standard. However, it could perhaps be considered eventually as an industry code of practice which could be used in a court of law as the accepted industry norm. When published in 1982 it was the only document of its kind, and it was agreed at the outset that since the state-of-the-art is so flexible, an authoritative document such as a Code of Practice may appear to be too restrictive in that the recommendations may have to be modified in the light of experience or with further developments in the technology.

The MTTA document describes the hazards involved in using robots, how to assess the risks and the basic methods of safeguarding. Emphasis is put on the use of safe systems of work particularly during robot programming and maintenance, and the need for all persons to be adequately trained is strongly emphasised. Items which were difficult to resolve included rules on software interlocking, and until more definitive experience is available it is recommended that software interlocks should not be used as primary safeguards without the backing of separate hardware interlocked systems. Requirements for controls received particular consideration, such as the need to distinguish between master and emergency stops, the use of two hands on teach pendants and a facility for slow speed operation, all of which are not included on many currently available robots.

The results of MTTA's work have been well received and proposals are in hand to prepare similar recommendations for specific robot applications such as welding and automatic assembly.

The principal standards-making body in the UK, the British Standards Institution (BSI), has recently formed an *ad hoc* committee responsible to Committee OIS/19, Industrial Automation. PEL/84 (Electrical Equipment of Machines), and MEE/6 (Machine Tools) also have an interest in this work. Safety considerations will take low priority initially, and terminology, classification and performance are the main topics for study.

As mentioned previously, the HSE has taken the lead in the UK in establishing its safety framework for risk assessment[2] and it is likely that this will form the basis of many future robot safety standards. Essentially the framework considers each mode of operation of the robot. A hazard analysis is then carried out for each mode and under 'designed' and 'aberrant' behaviour. If the hazards are liable to lead to injury then this determines the standard of guarding required.

European collaboration

It is interesting that the Safety Framework for Robots drawn up by the HSE has recently been amplified and is forming the basis of tripartite talks with West Germany and France. Meetings at government level between experts from the three countries were instigated early in 1981 and several working groups have since been formed, one of which is specifically concerned with robot safety. In a similar way to the original HSE safety framework and the MTTA guide, the proposed new framework will consider an assessment of the various modes of robot operation, the assessment of the hazards and the design and suitability of possible safeguards.

It is not yet clear what the ultimate outcome of this initiative will be – a European or an international standard or maybe a series of national standards using the same framework. Nevertheless the approach is welcome in that collaboration at this level between different countries can only be beneficial in the long term. Participation by other European manufacturers such as Sweden and Italy would also be welcomed in this work.

This involvement by the HSE in the European tripartite talks is being monitored by the new HSE Robot Safety Liaison Group with representation from UK robot manufacturers, users and research establishments.

Robot standardisation is also being considered by a new EEC working subgroup called Robotics Europe. This group, which comes under an 'informatics' EEC umbrella, is a forum on robotics standards which coordinates activities and provides funds to support them. Health and safety is included in its terms of reference, in addition to terminology, performance, etc.

International work

International cooperation on robotics safety standards is still in the planning stages. The International Standards Organisation Committee, ISO/TC 184, concerned with industrial automation, has formed a sub-committee on industrial robots and has produced a work plan for the establishment of standards on terminology, classification, performance and safety. A framework for a safety standard has been agreed and comparisons made of British, Japanese, German and American recommendations. A first draft should be available in 1985.

The International Electrotechnical Commission (IEC) has also recently formed a working group of TC44 (Electrical Equipment of Machines) to look at the safety aspects of programmable controllers for robots.

National standards

Fortunately fatalities with robots have been few although instances have been reported of near misses and aberrant robot behaviour. What has been recognised however, is the potential for injury and the need to restrict free access to robot installations. Many robot user countries have now established rules and standards for safeguarding robots and although there may be some criticism of overguarding in some cases it can be argued that it is better to be safe than sorry.

The initial work on preparing recommendations on the safeguarding of industrial robots was started in West Germany in 1979 with an early VDI draft and in the UK in 1980. As indicated earlier, the latter was issued as guidance by the MTTA in 1982[1] and was quickly followed by East German (similar to the VDI draft)[3] and Russian standards[4]. Japan issued a standard in 1983[5] following two fatalities and a government survey of robot accidents (see also page 23). Recently a revised and considerably enlarged draft of the VDI recommendation has been published[6] and drafts by Afnor in France[7] and by the Robotics Industries Association of America are in preparation[8].

All of these standards are recommendations rather than being obligatory. However in Russia and Japan they have led to amendments in legislation. In particular in Japan the law specifically now asks for measures to be taken to prevent hazards when robots are used and to educate robot operatives.

Considering that all the national standards have been produced independently it is surprising how similar the approaches have been in the different countries. This may be due not only to the early recognition of the hazards associated with industrial robots but also to the cooperation between the countries involved and their willingness to exchange information in the draft stages.

All of the national standards give recommendations on the design, construction and safeguarding of robots. Additionally the West German, American and British documents describe requirements for the safe installation and use of robots and give advice on the need for training and education of all personnel involved.

Surprisingly Sweden and Italy as major robot manufacturers have not produced any safety standards, although Swedish surveys on robot accidents have provided probably the best source of information (see page 49).

Design requirements

The design requirements include the elimination or enclosure of potentially dangerous parts, use of anti-overrun devices, layout of controls, power requirements including protection against irregularities in the power supply and electrical interference, and the need for the robot manufacturer to provide information.

It is the control system design which is currently providing most controversy from a safety point of view. Most robots are programmably controlled and if required safety can be built into the software of the program. Whilst all the standards ask for an emergency stop, so far it is only the British guidance which specifies that this should be hardwired (although the German document implies this by referring to other standards thus providing a higher level of safety integrity). This is particularly important on teach pendants which not only should be provided with an emergency stop but also have an automatic slow-speed facility. In the Japanese and Russian standards this is spelt out as 30 cm/s. The German and American drafts prefer 25 cm/s but these speeds appear to be based on current practice rather than any criterion which implies that speeds below this figure are safe.

Safeguarding

All of the standards state the requirement of preventing access to the robot when the automatic cycle is capable of being initiated. In general this means a guard around the robot which is interlocked to the robot cycle. In West Germany, France and the UK this usually means fencing 1.5–2 m high. In Japan a lower standard seems to be accepted with single or double rail barriers. The British approach points out that the type of guarding depends on the application. They recommend that a risk assessment is carried out and that as a result guarding may be not required on, for example, small robots. The UK document also emphasises the dangers of using software interlocking for guards unless redundancy techniques or separate conventional hardware interlocks are used.

User requirements

The American and British robot safety guidelines each outline the need for safe working procedures, in addition to mechanical and electrical safeguarding. These include the use of work-permits and rules of access within the guarding for programming, teaching and maintenance. The German draft also gives rules for robot operation, both in automatic and setting-up modes.

Practical experience with industrial robots has shown that programming and teaching is an important aspect of robot safety. The need in many cases for the programmer to be close to the robot arm, the possibilities of corruption of programs through electrical interference, or human carelessness, can lead to hazardous situations. Recommendations are given in the British, American and German documents.

Little reference has been made so far to the Russian standard which may have been based on the early Japanese and German drafts. One interesting aspect here is a requirement for the user to record and register any mishaps which occur. This, if treated seriously, could be a useful inclusion in any international standard, as data on robot accidents and, in particular, near misses, is often difficult to obtain.

References

1. Machine Tool Trades Association, 1982. *Safeguarding Industrial Robots,* Part 1: *Basic Principles.* MTTA, London.
2. Barrett, R.J., Bell, R. and Hodson, P.H. 1981. Planning for robot installation and maintenance: A safety framework. In, *Proc. 4th British Robot Association Annual Conference,* 18–21 May 1981, Brighton, UK, pp. 13–28. British Robot Association, Bedford, UK.
3. East Germany, TGL 30267/01, 1982. *Industrial Robot for Machine Tools; Terms; Requirements, Safeguarding Measures.*
4. USSR, GOST-SSBT, 1982. *Industrial Robots, Robotised Installations and Robotised Shops.*
5. Japan, JIS B 8433, 1983. *General Code for Safety of Industrial Robots.*
6. West Germany, VDI Guideline 2853, 1984. *Safety Requirements Relating to the Construction, Equipment and Operation of Industrial Robots and Associated Devices.*
7. France, AFNOR Standard, in preparation.
8. USA, Robotics Industries Association, 1984. *Robot Safety Guidelines.*

Systematic Robot-Related Accidents and Standardisation of Safety Measures

N. Sugimoto

*Research Institute of Industrial Safety,
Ministry of Labour, Japan*

Great hopes are entertained that industrial robots will prevent industrial accidents. But new forms of industrial accidents due to robots have occurred and have been reported. Since 1973, in Japan, JIRA, MITI, the Ministry of Labour and other industrial associations have formed committees and programmes for investigating safety measures and safety functions concerning the prevention of the hazards involved in robot operation. Most significant was the nationwide fact-finding robot-accident survey conducted by the Ministry of Labour. The results and amendments to the Ordinance are summarised.

Remarkable progress has been made recently in production technology. Worthy of particular attention is the vigorous development and introduction into the private sector of automatic production facilities, including industrial robots. Industrial robots, with their inherent flexibility, are demonstrating capability in a wide variety of manufacturing systems. It is expected that future efforts toward automation of a manufacturing process will focus on robotics at an accelerated rate.

As far as accident prevention is concerned, robotic techniques are most useful in dangerous or harmful environments where it has been heretofore believed too difficult to find any effective countermeasures. However, new forms of industrial accidents due to industrial robots have occurred. Safety, therefore, is one of the important factors to be considered in coping with an expected increase in the use of industrial robots.

Many studies have been conducted on the countermeasures for safety problems related to industrial robots. The standardised safety measures and safety functions that the manufacturers of industrial robots should provide for their machines were reflected in the Japan Industrial Standard JIS B 8433 (the General Standard of Safety of Industrial Robots). Furthermore, procedures for the safe use of industrial robots were reflected in amendments to the Ordinance on Industrial Safety and Health and the establishment of technical guidelines for industrial robots.

Survey of robot-related accidents

This survey was conducted by the Prefectural Labour Standard Offices of the Ministry of Labour in July 1982, with a view to looking into how industrial robots are actually used in manufacturing industry, as well as to identifying problems concerning safety and health involved in operations using industrial robots. The survey was conducted on 190 industrial plants, where a total of 4341 robots were in use.

Of all workplaces surveyed, 60.5% had introduced industrial robots for the first time since 1980, to eliminate dangerous work (55.3%), to eliminate harmful work (58.9%), to mechanise laborious work (45.3%), and to mechanise monotonous work (69.5%). In terms of robot type, these workplaces were operating: fixed sequence robots (42.4%), variable sequence robots (9.6%), playback robots (39.0%), NC robots (7.6%), and intelligent robots (1.5%). Applications were for welding (41.4%), workpiece loading and unloading (36.4%), painting (3.7%), assembly (12.4%), inspection (7.7%), others (3.7%). Of the workplaces surveyed 83.2% were planning to introduce additional industrial robots within five years.

As far as safety measures are concerned, 89.5% of the workplaces surveyed had taken some measures to prevent hazards resulting from contact with robots. Such measures included: enclosures (railings, fences, etc.) (66.5%), rope, chains, etc. (23.5%), 'keep out' signs (5.9%), and light-ray system safety devices and ultrasonic sensors (4.1%). An emergency stop is fitted to robots in 95.8% of cases, and 93.4% of these devices had resetting functions. Workplaces where periodical inspections and repairs were carried out over the past year accounted for 67.4% and 57.9% of the total workplaces surveyed, respectively. Furthermore, 89.5% carried out inspection before the robot started working. Also, training measures were undertaken for those workers involved in teaching and adjustment, operation and supervision, periodical inspections, and repairs, at 90.9%, 82.6%, 66.3% and 63.2% of the workplaces surveyed, respectively.

The total number of industrial accidents due to robots at the 190 workplaces surveyed was 11 (2 fatalities, 2 injuries leading to sick leave, and 7 injuries with no work absence). Of these, seven cases occurred after 1981 (see Table 1). Also, 37 near-accidents were reported of which 33 cases took place after 1981.

Table 1   Number of accidents by year of occurrence

Year	Accidents	Near-accidents
1978	2	2
1979	2	–
1980	–	2
1981	6	13
1982	1	20
Total	11	37

Accident patterns

It is very difficult to standardise safety measures because robots are used in different ways, and because information on accidents has not yet been systematically documented. However, by analysing the many incidents recorded, the Research Institute of Industrial Safety has extracted some typical patterns that occur in robot accidents. Eight of the 11 accidents (73%) were due to the unexpected movements of robots, whereby the person entered the work area when the robot had stopped working (or was working very slowly) and the robot suddenly and unexpectedly started moving (or accelerated).

Fig. 1 is a fault-tree analysis of results ordered according to the element of the unexpected robot movement. Mishandling accidents such as scattering or falling of materials are also included since there are some similarities. To prevent accidents caused by unexpected robot movements, we have to find where and how often these potential dangers appear. Again, using data obtained from the Ministry of Labour investigation and also from data of 350 cases of incidents related to industrial robots.

Table 2 shows that 28.3% of robot-related accidents were caused by the unexpected 'start-moving' of the robot manipulator which apparently had

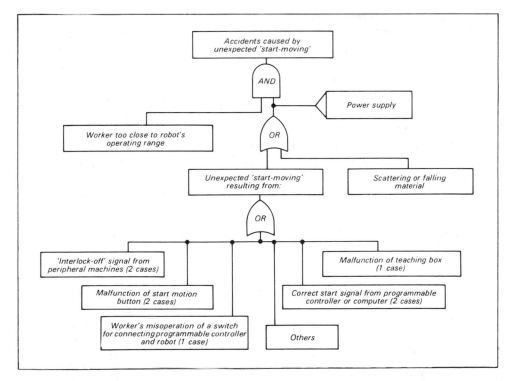

Fig. 1   Fault-tree analysis of results ordered according to the element of the unexpected robot movement

Table 2 Relation between accidents and
movement of manipulators

Type of movement	Percentage
Unnatural movement	30.6
Unexpected movement	28.3
Stops moving suddenly	5.26
Does not start to move	9.54
Not concerned with movement	17.4
Scattering or falling of materials	8.5

stopped moving. Adding to this accidents caused by scattering and/or falling materials, the figure becomes 37.2%, thus illustrating the potentially high danger if a worker carelessly gets too close to the manipulator.

Table 3 classifies the accidents and shows the causes of different possible unexpected 'start-movements': 38.1% of the accidents were due to erroneous operator actions whereas 61.9% were caused by robots. However, the latter includes faults which occurred in peripheral machinery which lead the operator to the danger area for work-handling or trouble-shooting purposes.

Establishment of a technical guideline for accident prevention

The Japanese Industrial Safety and Health Law Article 20 requires the employer to take necessary measures to prevent hazards that may arise from the operation of machines, equipment and other facilities. These necessary measures are concretely specified in the Ordinance on Industrial Safety and Health enacted in accordance with the said Law.

Furthermore, it is stipulated in Article 28, that the Ministry of Labour shall make public technical guidelines for any kind of industry or any type of work to ensure the appropriate and effective implementation of measures which the employer is required to take pursuant to the said provisions.

The industrial robot is fitted with a manipulator which performs functions equivalent to human hands. This manipulator characteristically moves automatically in space outside the machine body, but the direction,

Table 3 Classification of accidents and causes of different possible unexpected 'start movements'

Classification of accidents	Percentage	Formation of the 'unexpected start-moving'
Breakdown of an electric part (controller, etc.)	52.2	23.9
Defect of a mechanical part	8.5	5.3
Breakdown/defect of the actuator	7.16	6.19
Misconnection of the peripheral machine	3.75	8.0
Disorder of air, oil, electric source	1.71	0.9
Scattering or falling of materials	7.51	16.8
Erroneous action by operator	18.4	38.1
Others	0.88	0.81

operational speed and sequence in which it moves cannot be predicted easily. Most of the industrial accidents due to robots resulted from workers coming into contact with a very powerful manipulator having such characteristics. In some cases teaching, which is an operation peculiar to robots, is conducted by a worker standing close to a moving manipulator. Thus, high priority should be given to safety in operations of this sort.

Since existing machinery did not have these characteristics, there was no relevant provision in the Ordinance on Industrial Safety and Health. Furthermore, now that robots arc highly complicated and advanced in construction and control, they require a considerable knowledge in their operation. Lack of this knowledge will no doubt result in industrial accidents.

With these points in mind, the Ordinance on Industrial Safety and Health was recently revised to incorporate necessary provisions for industrial robots and a partial amendment was made to the Rules of Special Education on Safety and Health (July 1983). In addition, complementary to the Ordinance on Industrial Safety and Health, a technical guideline was newly established (September 1983).

Amendments to the Ordinance

The Ordinance was amended to control industrial robots, except for those under research and development and those specified by the Minsitry of Labour, which are equipped with manipulators and memory systems (including variable and fixed sequence control systems) and which allow manipulators to expand and contract, bend and stretch, move up and down, move from side to side, circle round, or perform a combination of these motions automatically in response to information stored in memory systems (related to item 31 of Article 36).

The Ordinance does not apply to small robots such as those which have an operating range limited within a cylindrical space of 300mm radius and a height of 300mm (cylindrical type), or within a spherical space with radius of 300mm (polar type or joint type), or those which have the operation distance of 300mm or less (Cartesian type), or those driven by an electric motor with the related power of 80W or less per moving joint.

Robots in operation

When workers intend to operate an industrial robot and if there is a possibility of danger as a result of their coming into contact with the robot in operation, the employer is now required to take the necessary measures to prevent such hazards by setting up a railing or an enclosure (related to Article 150-4); providing that this provision does not apply to cases where the following measures have been taken to prevent hazards due to the unexpected entrance of workers into the operating range of an industrial robot:

- Sensor technology to detect workers entering the operating range, such as photoelectric beams, safety mats, ultrasonic, capacitance, microwave, infrared and vibration-sensors, and camera surveillance.
- A rope or a chain enclosing the operating range with an 'at work' sign.

- Supervisors who check entry of personnel into the robot's operating range.
- A camera surveillance system that enables the supervisor to stop the robot immediately.

Teaching

The necessary provisions concerning the prevention of hazards involved in teaching, and other operations, were also incorporated. The employer is now required, in principle, to take the following measures (related to Article 150-3):

- To lay down rules for specific matters concerning the industrial robot, including its operating method and procedure, and to let workers perform operations in accordance with such rules.
- To ensure that operation of the robot can be stopped immediately on finding any abnormality.
- To mark the starting switch, etc. of the robot with an 'at work' sign.

When workers perform teaching and other operations within the robot's operating range, the employer is now required, in principle, to inspect for specific matters, including functioning of an emergency stop device before starting operations, and if any abnormality is found, to take the necessary corrective measures (related to Article 151).

Inspection, repair, maintenance

It is now stipulated that the employer ensures that the robot is stopped before workers perform inspection, repair, adjustment, cleaning and lubrication within its operating range. While such operations are being performed, the robot's starting switch must be locked in the 'off' position, except when these operations have to be performed while the robot is moving. The above three measures as stipulated for 'teaching' also apply (related to Article 150-5).

Safety education

Special education on safety is now mandatory under the Ordinance. It should be given in respect of the following additional operations (related to items 31 and 32 of Article 36):

- Teaching and other specified operations to be performed within the robot's operating range.
- Operations to be performed while the robot is moving, including inspections, repair and maintenance.

Technical Guideline

The newly established Technical Guideline outlines the points to be considered in selecting industrial robots, measures to be taken in their use, details of periodical inspections, and methods of education for workers in order to prevent industrial accidents.

The Guideline consists of the following items:

- *General provisions* – Purpose, definition, application exemption.
- *Selection* –(a) Construction: emergency stop, safety functions, control

panel, input/output terminals, grippers, 'workability' in teaching and other operations, projections on outer surface, release of residual driving pressure, indication of manipulator's direction of movement; (b) adaptation to environmental conditions; (c) marking; (d) instruction manual.

- *Installation* – Layout, stopper, confirmation of operation.
- *Use* – (a) Measures for prevention of contact: enclosure, light-beam safety device, rope or chain, supervisor; (b) measures relating to tasks within the robot's operating range: operation roles, marking on control panel, inspection before teaching and other operations, cleaning tools, release of residual pressure, test operation, luminous intensity; (c) measures taken during operation of robot: starting, emergency stop; (d) prevention of 'flying' objects released by robot.
- *Periodical inspection* – Before commencement of work, repairs.
- *Education* – Details of education, persons in charge of education, emergency measures.
- *Miscellaneous* – Control of magnetic tapes, etc.

Recent activity

Over 500,000 workers in Japan are engaged in jobs directly or indirectly related to about 150,000 robots. They are now allowed to teach or inspect their robots in motion within the operating range without acquiring the special education as the Ordinance prescribes.

To ensure 'robot safety', it is important to fully understand the industrial robot itself and related matters. The Ministry of Labour provides a 'special education' textbook which covers a wide range of knowledge on the structure and control of industrial robots and information on teaching, maintenance, repair and other robot-related operations. In addition, the Ministry conducts a training programme for instructors who are in charge of education in industrial safety. Over 1000 instructors have completed the 40 hours education and training. Consequently, many workers have completed the special education required and obtained qualifications to teach, maintain and repair industrial robots and other robot-related operations.

Industrial robot manufacturers have endeavoured to develop their machines to satisfy the JIS and the Guideline. Indeed, many industrial robots have reached the requirements of both documents. The JIS and the Guideline do not deal with safety measures of design on whole systems of robots and peripheral equipment, as it is difficult to regulate in a standardised way because of the large range of systems both in terms of application and scale.

The Research Institute of Industrial Safety has recently been given a US$2 million budget by the Ministry of Labour (fiscal year 1983), for research into safety problems and countermeasures related to the automated production system in close cooperation with manufacturers, users and system analysts/designers of industrial robots, other automated machinery, factory automation, FMS, etc.

2
Surveys and Analyses

This section includes work from West Germany, Sweden, the UK and Japan. The first paper examines the problem of occupational safety from a broad viewpoint showing how accidents could arise, how hazard analysis is performed, and how checklists, flowcharts, etc. can help. Nicolaisen provides a wealth of information which will be helpful to the robot workplace designer, and a summary of some of the accident statistics from a 14-day investigation in Sweden.

The next paper looks at the Swedish accident statistics during the period 1979–83. Each of the 36 accidents which occurred in that period are clearly described. It is interesting that of the 36 accidents 12 occurred during planned or unplanned interruptions. Also, of the 36 accidents, 6 were head injuries and so were potentially very serious. Carlsson concludes with a set of risk elimination and risk reducing measures.

In contrast, the next paper by Jones and Dawson is an analysis of 37 robot systems in three companies involving 20,000 robot production hours. Among the interesting points to note from this study is how much is not known, and the difficulties in recording and properly attributing failures.

The final paper in this section by Sugimoto and Kawaguchi provides a useful indicator of the various things that can go wrong with the total system. Although the detail of the fault tree and the statistics of types of failure and mean time between failures will eventually need updating, the concepts illustrated are more permanent and are extremely important for designers and for the analysis of potential problems.

Occupational Safety and Industrial Robots

P. Nicolaisen
Fraunhofer-Institut für Produktionstechnik und Automatisierung
(IPA), West Germany

Often, when industrial robots are in use, little attention is paid to occupational safety. However, it is very difficult for many users to find a suitable solution for their problem. Beginning with a representation of the problem spectrum as a whole, starting points are indicated and discussed for the improvement of safety at industrial robot workplaces.

Industrial robots have high power movements and are freely programmable with regard to route and speed of movement. Accident hazards arise because, even when behaving as planned, it is not usually possible for an outsider to predict the next movement. Errors, for example, in the positioning control or in the speed monitoring, can produce completely unpredictable movements with undefined speed within the kinematically possible range of movement.

With conventional machines movement usually takes place within the machine, whereas with industrial robots the size and shape of the danger area, i.e. of the relevant working space and maximum range of movement, respectively, is not immediately recognisable. For this reason precautions must be taken to safeguard this zone[1].

The accident event

Most people who have dealt with industrial robots can remember critical situations in which only good fortune has prevented a 'near' accident from becoming an 'actual' accident. It is, however, difficult to find statistics on accidents. Most of the countries using industrial robots have either no data or only vague data available, and the percentage of unrecorded accidents could presumably be high.

Sweden, however, forms an exception in this respect[23]. (See also page 49). The results of a fourteen day survey of eight industrial robot workplaces are given in Fig. 1. In the course of this investigation, dating from 1981, a total of 24

Dangerous situation observed	Cause	Pointers for solving the problem
Robot brushes against operator	Control panel is within the enclosure; operator passes through the zone where the robot is moving	Layout
Robot brushes against person	No all-round enclosure; there is a gap between two machines	Layout
Robot brushes against person/operator/machine-setter during his work	Parameters of machine are adjusted and material fed in while the plant is running	Layout, organisation of work
Operator is struck by the robot	Sometimes the workpiece remains hanging on the conveyor belt and the operator intervenes while the installation is moving	Work organisation, layout, working process
Operator is struck by robot while he is working	Range of activity (workpiece being put from bunker into store) is partly outside the working zone of the robot	Layout, work organisation
Robot projects the workpiece at high speed	No pressure or too low a pressure for pneumatic gripper	Design, layout
Robot brushes against person (toolsetter)	Pneumatic cylinders have to be adjusted with the installation in operation	Layout, work organisation
Danger of being burnt by HF installation	There are various emergency cut-out circuits; if the emergency cut-out circuit is released the HF preheating system starts automatically	Layout, (interlinking)
Person is struck by arm of robot or by workpiece	Safety grid fitted wrongly and is too low; robot reaches out over the fence	Layout
Person is struck by robot while he is working and can be burnt by fluid	Range of activity (feeding in liquid metal) is partly within the programmed movement of the robot	Layout, work organisation
Person is struck by robot, danger of burns from liquid spray	Remains of cast metal are usually removed from the workpiece while the installation is running	Layout, work organisation, working process
An inquisitive person is struck by robot, danger of burning from hot workpiece	Gaps in fencing	Layout
Hand of operator gets jammed between robot and barrier *Injury (7 days absence from work)	Too short a distance between robot working zone and barrier Barriers mesh is too wide and/or barrier is too low	Layout
Programmer is struck by robot although he is acting correctly	When, during the manual type operation (programming) the safety barrier is opened, the robot does not stop immediately but completes its cycle	Layout, interlinking

Fig. 1   Accident hazards with the use of industrial robots – results of an observation at eight workplaces over a period of 14 days

Dangerous situation observed	Cause	Pointers for solving the problem
Person is struck by robot	Safety barrier has gaps because after a machine has been replaced the old (now too small) barrier is still in use	Layout, work organisation
	Palletising system for workpieces is part of the safety enclosure; when pallets are removed there is free access	Layout
Operator is struck by robot	Operator, in the course of his work, (lifting the workpiece and inserting it) bend into the operating zone of the robot	Layout, work organisation
Slipping on oily floor	Because the spray nozzle has been wrongly positioned oil runs over the robot axis and onto the floor behind	Working process
Operator's hand is jammed between the robot and hot workpiece *Burn injury	The amount of oil passing through the robot is sometimes insufficient (because of faulty nozzle setting) so that when the plant is running more oil has to be added by hand	Working process
Operator is struck by blown-out metal pieces (trimming press) *Cuts due to metal splinters but no actual accident involving robot	Different reasons: position and pressure of blow-out nozzles, position of operator	Layout, work organisation, working process
Stumbling over cable between robot and control panel *Injury (bruising)	Cable not covered and badly laid	Layout, work organisation
Maintenance worker is pressed by the robot against a running grinding wheel *Cuts and burns (14 days off work)	Maintenance worker had stopped the plant via the emergency cut-out but did not know that the emergency cut-out only stops the robot (hydraulics out). When, in cleaning, he inadvertently touches a terminal switch, the grinding machine starts up and the robot, without pressure, caves in and presses the operators arm against the moving grinding machines	Design, layout, interlinking
Risk of injury by projected workpiece	The feed line of a compressed-air operated gripping tool is broken off so that the workpiece can no longer be held	Design, layout
Robot makes uncontrolled movements and collides with machine, whereby the gripping tool is bent	A lead was broken by dangling workpieces and passed across the various control signals (including the emergency cut-out). This caused the entire plant, including the robot to get out of control	Design, layout, interlinking

*Injuries or material damage

critical situations were recorded, the cases indicated by an asterisk leading to injuries or material damage.

Potentially, the increasing range of applications of industrial robots will create increased accident hazards for two main reasons. First, owing to the wider use of small- and medium-sized mass-produced machines, the volume of programming work to be carried out will increase, implying more frequent contact between the person and the industrial robot. Secondly, the range of functions will be extended so that the manufacturing system as a whole becomes more complex and thus, in certain circumstances, becomes more prone to breakdown. Complexity arises from:

- more complex programs,
- collaboration of several industrial robots,
- use of sensors (external data presentation),
- use of gripper/tool changing systems, and
- mobile robots (e.g. mounted on vehicles).

In addition, the energy potential available within the system increases through the use of:

- industrial robots with higher performance,
- high-speed tools, and
- laser and water-jet cutting with industrial robots.

Present state of safety technology

So far, in spite of differently orientated problems, conventional safety technology has been used. The safety fence (Fig. 2), frequently designed so that when it is opened the installation automatically stops, is by far the most widely used safety device. The reasons for this are obvious:

- it is usually simple to produce,
- it provides inexpensive protection outside the fencing, and
- it provides protection against 'thrown away' parts.

So, for the near future, the safety fence will continue to be used. However, it is no universal panacea; there are situations where it cannot be installed because of the manufacturing process. Also, the safety fence affords no protection when work has to be carried out inside the enclosure (fitting, programming, maintenance, inspection, repair).

In practice, however, there is often a shortage of safety devices. This is primarily due to the following reasons:

- In the development and introduction of new technologies, safety problems tend to have a lower priority than questions of technology and sales policy.
- Awareness of a problem often develops only after several negative experiences, i.e. on the basis of a large number of applications and after years of use.

Fig. 2   Typical safety arrangements for an industrial robot workplace

- The assessment of the level of safety varies according to the interests of the manufacturer, user and safety authorities.
- The solution of the problem raises difficulties, as specially prepared information material and relevant aids such as checklists, catalogues, collections of examples and suitable safety devices are lacking.

Safety improvement

From the foregoing remarks it can be seen that there are two important sets of factors involved to improve safety at workplaces where industrial robots are in use.

First, in order to achieve success, it is essential to make all people concerned aware of the problem, for only then will safety problems be given the same priority as questions of technology or economy. Safety factors should be included in the considerations at a very early stage. Secondly, this awareness and motivation needs to be accompanied by parallel activities dealing with the solution of the problem, that is:

- proposing methods of finding a solution,
- putting forward and evaluating principles and examples, and
- proceeding with further and new developments, in order to close gaps or eliminate faults.

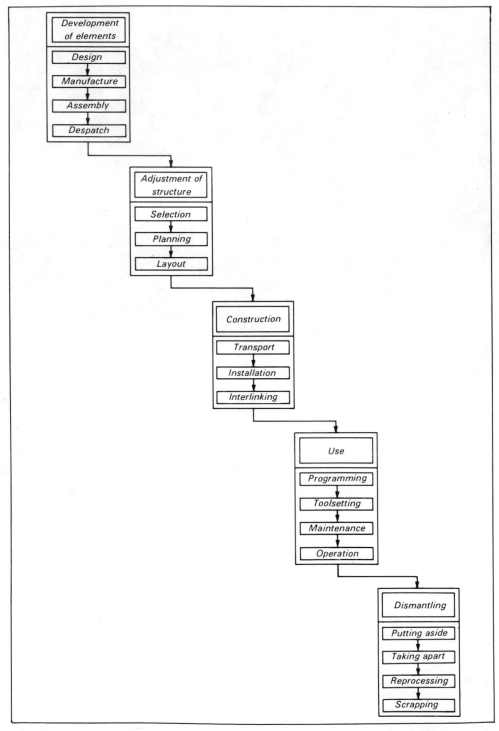

Fig. 3 Chronological development of a robot system-influence of prephases on the operating stage

- What hazards (especially new ones) arise in connection with robots?

- In which situations (activities) and at what places may there be a meeting of the person and the hazard?

- What possibilities are there for reducing the hazards and for avoiding the encounter of person and hazard?

- With what aberrations, that may become critical, must be expected in a system?

- What causes them (person, machine, surroundings)?

- When, where and how frequently do they occur?

- How are aberrations prevented?

- Is the system adjustable in the case of (all) aberrations?

- What possibilities of adjustment must be provided (possibility of communication in the case of aberrations)?

- How are non-reactions and/or wrong reactions prevented?

Fig. 4 Checklist of questions for accident prevention

Accident occurrence

A robot system consists of three elements: man, the industrial robot, and a communication system. Accidents can happen when hazards are present or occur, and also when men are present in the danger zone.

The stages in the development of a robot system are shown in Fig. 3. This shows that:

- A system is not static but undergoes development, whereby its elements stand in various relationships to each other and must function in different ways (e.g. installation–maintenance–dismantling).
- Before the most important part of a system, i.e. the application (operation), is reached, important preliminary decisions have already been taken with regard to safety (e.g. safe design–layout–maintenance).

In order to reduce the chance of accidents a checklist of questions, such as those shown in Fig. 4, should be asked.

The stages in an accident prevention procedure may be represented in a general way in the flow chart shown in Fig. 5.

Procedure

For the various stages in the 'workplace with individual robot' system, such as transport, installation/linkage, programming/arranging, etc., hazard analyses are carried out (corresponding with varying degrees of precision to the planning phase). The aim of these is to determine the existing dangers so that countermeasures can be taken.

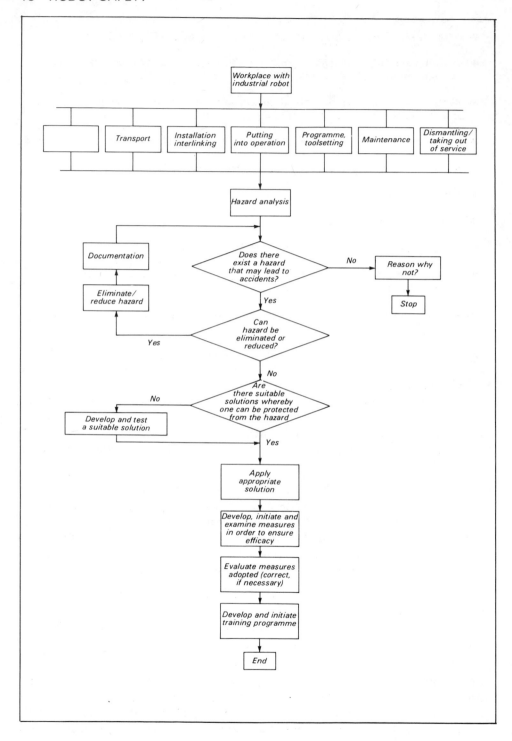

Fig. 5   Flow chart showing the different stages in an accident prevention procedure

The optimum moment for a first run-through of the flow chart is generally at a very early stage, for often, by fixing the plan or design (e.g. what type of control, drives with/without additional brake), we unknowingly take a decision as to the kind of safety that can be achieved for the system as a whole.

At it cannot be assumed that there have already been detailed discussions with experts from the occupational safety department (however desirable that would be), it is all the more important to acquaint the constructor and the planner of the plant with the fundamental ideas of occupational safety.

Aids

It is desirable to collect material and data relevant to the categories of problems shown in Fig. 6. Much thought should be given to defining, delimiting and solving the problems. This can be made more efficient by means of checklists, forms catalogues and sample solutions. A lot of activity is in progress, i.e. collecting information and exchanging experiences, within various groups. Although isolated, preliminary results are available and it is argued that the work should be coordinated, both in order to avoid duplication and in order to prevent each group working for itself and trying to impose its own method, thus building up insurmountable obstacles for its neighbours. One working group in particular has made this coordination its special concern[4].

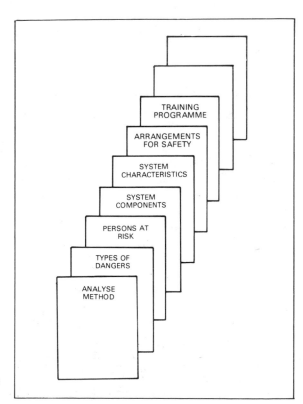

Fig. 6 Material and data collection categories

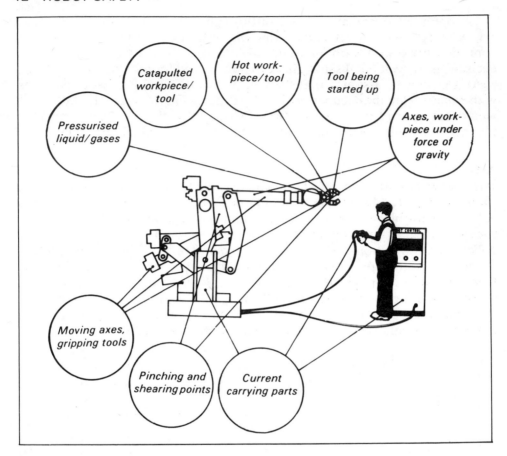

Fig. 7   Type and location of hazards when robots are used

The form and content of material and data collection, and also its use, can be best explained by an example, although it must be pointed out that at present only some of the pieces of the mosaic as a whole are present. The basis is a simple man–robot system. From the spectrum of the various system groups and functions the phase 'programming' is selected. Referring to the flow chart already shown we must first of all deal with the question of 'hazards' in the system. The first indication is given by graphs, tables, or collections of examples (which, of course, must be extended and made more manageable) as shown in Figs. 7, 8 and 9.

A step is advanced when, in the next run-through, the operational spectrum in the case of programming is also considered. It then becomes possible to focus on the main problems to find concrete pointers as to what measures should be taken.

Programming initiation
   Who programs?
   How are unauthorised persons prevented from intervening?

Type of hazard (type of energy)	Location of hazard (energy transmitter)	
	Specific to design of apparatus	Specific to use of apparatus
Kinetic energy	Moving mechanical elements –axes –cylinders –levers –chains –gearwheels –spindles –gear belts	Moving mechanical elements –grilling devices (open/closed) –tools/rotary (moving back and forward) –projected workpieces, tools or parts of these
	Flowing liquid/gases –driving materials (compressed air, hydraulic fluids) –auxiliary substances (greases, cooling media)	Pressurised liquids, gases –materials (paints, PVC, plasma) –driving substances (compressed air, hydraulic fluid) auxiliary substances (greasy, cooling, insulating media)
Potential energy	Axes descending by force of gravity Accumulators	Workpiece and tools falling by force of gravity Accumulators
Chemical energy	Hot, corrosive and toxic materials –driving substances –auxiliary materials	Hot, corrosive and toxic materials –workpieces, tools –driving substances –auxiliary materials
Electromagnetic energy	Current-carrying parts –drives –conductors –parts of mains –accumulators	Current-carrying parts –driving mechanism for tool transformer/electrode when welding
Infrared/lights/ uv-rays	Heat radiation emitting machine parts –drives –brakes	Heat radiation emitting objects –hot workpieces –welding electrodes
x-rays		x-radiation emitting objects –measuring equipment –testing equipment
γ-rays		γ-radiation emitting objects –radioactive workpieces –radioactive waste

Fig. 8   Hazards specific to design and application

How many are taking part?
Where and with what is it carried out?
What operational faults must be allowed for?
What effects do operational faults have?

Fig. 9   Examples of hazards associated with the use of industrial robots: (a) power of movement, (b) sharp edges (pinching and shearing points), (c) influence of gravity, (d) tool start-up (milling cutter), (e) hot tool, and (f) workpiece thrown out

Position of the programmer
Is there a safe position provided?
Is the position marked?
Is there means of preventing unsafe positions from being adopted?
Is it ensured that no other persons put the programmer at risk?
Must the programmer shift about between various positions?
Stance of the body at the position?
Visibility/light conditions?
Stress factors (noise, dust, heat)?

Reference point, transformation of coordinates
   Are conversion aids available?
   Is there unambiguity independently of position?
   What maximum distance (between man, reference point and object) is possible?

Control elements, selection of axis, selection of direction
   Arrangement of control elements?
   Type of control elements?
   Sequence?
   Functions realised?
   Control of operation?
   Marking of control elements?
   Marking of axes?   '
   Marking of direction?

In the search for a suitable solution, it is often helpful to consider different viewpoints:

- Accountability
  −manufacturer
  −user
- Category of measures
  −technical (safety technology)
  −organisational (organisation of procedure)
  −psychological (behaviour, motivation)
- Safety principles
  −eliminate/reduce hazard
  −separate man/hazard
  −prevent abnormality
- Subject of the measure
  −system element
  −system structure
  −points of intersection
- Risk category
  −kinetic energy
  −force of gravity
  −stored energy
  −electrical energy
  −radiation
  −chemical energy

Accident prevention measures during programming
- Reduce speed (very slow pace)
- Low surface pressure in the case of collisions (e.g. no sharp edges/corners, no projecting parts)
- Keep volumes for movements small
  −efficient layout planning

   –robot kinematics with satisfactory ratio of workspace to total range of
   movement
   –range of movement appropriate to function
- Rapid means of stopping in the case of a collision hazard
   –sufficient number of emergency cut-outs provided (also on teach-box)
   –additional brakes
   –deadman's switch
   –automatic stopping in the case of collisions (Fig. 10)
- Low probability of the robot not working properly
   –control: tested software; reliable hardware; controls for comprehensi-
   bility
   –drive: brake action even in the case of a power failure, and even under the
   effect of gravity; no automatic starting when power is restored
   –grippers: adequate solidity; adequate gripping force; no release if power
   is reduced/cut off
   –programming unit: adequate solidity; protection against environmental
   influences (dust, heat, vibration); no accidental operation (falling down,
   collision)
   –cable leads: adequate solidity; monitoring in the case of short-circuit and
   breakage
- Low probability of human failure
   –adequate training
   –comprehensible operational instructions

Fig. 10   Large-area safety switching device developed at IPA

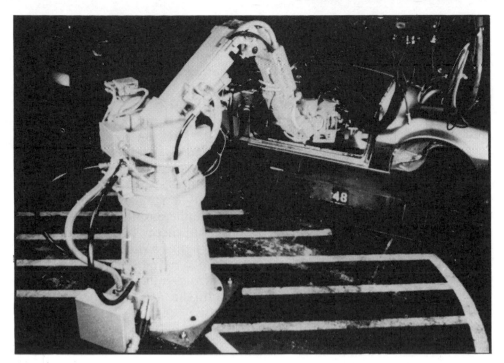

Fig. 11   Marking of danger zones

   –no interference by unauthorised persons
   –ergonomic design of operating unit
   –operator guidance
   –checking of decision instructions
   –no disturbing environmental influences
   –marking of the danger zones (Fig. 11)
● Sufficient distance between programmer and robot
   –programming from outside the safety enclosure
   –use of a programming stand

Although the collection of data and material on robotic safety has made progress, it has still a long way to go before it provides a real support for the design engineer, the layout planner or the safety engineer.

Other topics also need to be treated further. First, rules and guidelines for the design need to be developed. This is a wide field of activity because of the great number of possible factors of influence. The necessity to work out proposals for a safety-orientated layout-design requires, on the one hand, an analysis of accidents which have already occurred (more than 50% of all accidents are largely due to inadequate layout-design), and on the other hand on the knowledge that this problem will still exist even if tremendous progress is achieved in other areas, e.g. safety-orientated design of industrial robots. (A safe robot does not ensure a safe plant even if the robot represents an important factor of influence.)

Secondly, safety devices for robot workplaces need to be investigated. A catalogue containing all available safety devices has to be compiled, complemented by a collection of examples of all realised solutions. It may be necessary to improve existing or to develop new safety devices to achieve better solutions to the problems. A step in this direction is the previously mentioned large-area switch developed at IPA (Fig. 10); but besides its use in robot workplaces there is still a wide range of possible applications in industry (e.g. guided vehicles, moving parts of machines). This device offers effective protection especially for operations in the immediate vicinity of the robot, since it is fitted directly on the apparatus and travels with it. The device consists, essentially, of two components: a large-area switch, which on contact with an obstacle (person, machine, support) gives a signal which stops the movement; and a suitable structure which, like the dashboard in a car, can change on impact and thus contain the motional energy still remaining after the disconnection. The signals given by the large-area switch are evaluated by means of an electronic circuit, which according to the requirements for safety devices of the German Safety Authorities, controls faults such as the breakage of cables or short-circuits, and helps to maintain the safe function of the robot in the event of component breakdown.

References

1. Nicolaisen, P. 1980. Problems of occupational safety with the use of industrial robots. *wt. Zeitschrift fur Industrielle Fertigung,* 70 (1).
2. Arbetarskyddsstyrelsen. *Accident Statistics of the Swedish Safety Authorities.* ASS, Solna, Sweden.
3. Work Environment Foundation. *Research Projects of the Swedish Work Environment Foundation.* ASF, Stockholm.
4. Barrett, J., Bell, R., Duelen, G. and Nicolaisen, P. 1981. Problems of occupational safety in connection with the use of industrial robots. Paper presented at the *Meeting of the British–German–French Working Group on Aspects of the Safety of Industrial Robots,* Paris.
5. Nicolaisen, P. 1980. The development of safety devices to fit the problems–Example: industrial robots. *Industrieroboter,* 102 (73).

Robot Accidents in Sweden

J. Carlsson
National Board of Occupational Safety and Health, Sweden

Accidents involving industrial robots are unusual occurrences. In the years 1979–1983, inclusive, there were 36 accidents in Sweden which resulted in human injury, thus averaging about eight cases a year. Circumstances of robot-related accidents are described together with actual descriptions of the accidents investigated.

This report is based on data obtained from the computer file ISA, the Information System for Occupational Injuries, which is maintained by the National Board of Occupational Safety and Health in Sweden.

Various occupational injury reports which make mention of industrial robots were identified as the chief cause of injury in 36 cases. Considering that occupational injury reports often lack sufficient information, this study may be marred by two flaws: some injuries were not recorded and some were not directly caused by robots. All 36 injuries are detailed in the appendix to this paper. Hence the reader is free to decide the validity of the conclusions drawn from the accidents reported.

Previous investigations

The first Swedish report was published in 1979[1]. Between 1976 and 1978, there were about seven accidents per annum. Thus, together with the findings of the present study, from 1976 to 1983 inclusive there were seven to eight cases each year of accidents involving industrial robots that required 'sick-listing' of the injured person.

The conclusions that have been drawn about robot installations and robot equipment remain valid and agree with this material. The report presents a listing of the safety criteria that govern the installation and operation of industrial robots. Most of these criteria accord with the requirements set forth in the National Board's then-governing instruction manual for machinery (1978). However, special stipulations have to be laid down for industrial robots, notably because the robot moves in a larger volume of space and because the robot arm can assume 'unexpected' positions.

Other reports set out safety measure proposals in robot installations[2] and patterns posed by accidents due to faults in sophisticated control systems[3].

Circumstances of robot-related accidents

This report is confined to accidents that occurred because of the injured person's contact with a robot while it was operating, or caused by the robot's special functions as per definition. Excluded from the material are accidents due to operator over exertion when repairing a robot or falling against a robot. In any case accidents of this nature are rare.

Types of robot

Manually controlled manipulators are the type most often causing accidents, whereby the manipulator is carrying out pick-and-place operations (Table 1). The other types of robot included in this study are those with tools such as spray guns for painting, butts for spot welding, and so on. In a few instances the functions were not clearly stated in the accident report form. (It is likely that four of the accidents listed in Table 1 under 'miscellaneous', were also caused by manually controlled manipulators.

Table 1 Number of accidents by robot type

Type of robot	No. of accidents
Manually controlled manipulator	28
Welding	1
Painting	1
Coil-winding	1
Miscellaneous, unclear	5

Types of event

Most of the accidents (14) occurred while adjustments were being made in the course of normal operation (Table 2). This means that the robot was functioning as intended but that the flow of materials had been disrupted making it necessary to adjust the robot manually. In almost as many cases (13) accidents were due to the robot making an unexpected movement while it was being programmed, repaired or the position of its arm changed. In a few cases the worker 'bumped' into the robot, which was not then in operation but in an unexpected position.

Judging from the accident reports, it is the manual adjustment of material in proximity to a robot arm, which has no protective screen or guard, that is the most common cause of injury. In other words, the injured person has been

Table 2 Number of accidents by event type and activity of the injured person

Activity	Contact with moving machine, part, material	Other
Adjustment in course of operation	14	—
Movement up against robot	1	1
Repair, programming, etc.	13	2
Miscellaneous	3	2

working inside the robot's work area. It is not common for injury to befall unauthorised persons (the study has only two such cases).

Misunderstandings

Six accidents were reported whereby the operator thought the robot had been switched off yet had come in contact with a moving part of the robot.

In two cases the injured person had inadvertently touched the robot's turn-on button and was injured by its arm.

In at least one instance the accident occurred because the operators failed to understand one another when performing adjustment/programming.

Industrial area

By far the principal venue of accidents was machine shops in the engineering industry. In any case most industrial robots are installed in such workplaces (Table 3).

Table 3 Number of accidents by industrial area

Industry	No. of accidents
Automotive	9
Other metalworking	16
Foundries	4
Plastics	4
Other	3

Robots are fairly common in the automotive industry. As a result, car plants are the places where most of the accidents with industrial robots occur (nine in this study).

Job title

The job title of the injured person will often give a pretty good idea of the kind of work involved in the robot environment. Is the injury of a casual nature, of the sort that is to be expected, or what? As it happens, a large number of the injured persons are repairmen, testers, and the like (Table 4).

One-third of the injured (12 in total) are persons who work with the machine in the course of planned or unplanned interruptions, or persons who instruct operators.

Table 4 Number of accidents by job title of injured persons

Job title	No. of accidents
Repairman, set-up man, mechanic, machine setter, foreman, instructor	12
Other job titles, operator	24

Types of injury

The principal injuries are inflicted on the hand, fingers and head, and usually entail short periods of sick leave (Table 5). Head wounds are fairly numerous, accounting for one-sixth of the injuries.

Table 5 Number of accidents by types of injury

Injured part of body	No. of accidents
Finger	12
Hand	7
Arm	2
Back	4
Head	6
Neck	1
Leg	2
Rib	1
Tooth	1

Concluding remarks

Robot-related accidents comprise a small portion of all accidents involving machines and devices. They do not require more than relatively brief periods of sick leave, judging from the data available for study. However, many injuries were head wounds, and these can be very serious indeed. So there is every reason to bear such hazards very much in mind when robots are to be installed. A sizeable portion of the accidents can be explained by the presence of the injured person inside the robot's work area.

The following measures serve to eliminate risks:

- The separation of operator workplace from robot workplace is of great importance.
- The robot's work area must be screened off by means of automatic shutdown devices.
- A robot must be programmed and test-run from outside its work area. If that is impractical, the robot will have to be tested at reduced power and creeping speed.
- No repairs are to be made on a robot until the energy source has been removed.

Since risk-eliminating measures are often difficult to implement, the following risk-reducing measures are worth considering:

- Equip the robot with stopping gear, which shuts down the motion of the robot arm/tool whenever there is physical contact with a person.
- Paint the robot arm in a special colour to make its position clearly visible.
- Design the robot to reduce the risk of cuts and pressure contusions.
- Place the robot where no one can get squeezed between parts of the robot and stationary objects in the vicinity.

- Whenever the robot works in tandem with another machine/device, it should be possible to shut off both systems at the same time with the same stopping gear.
- Sometimes, when a robot is to be programmed or adjusted, some person will be engaged in setting the robot's positions. Another person will be stationed next to the robot tool in order to control these positions. In situations of this kind, both persons should have access to the stopping gear.
- In case of emergency, the robot arm should have its power cut off, but should stay where it is so that the person who has caused the shutdown will not be struck by the robot arm.

References

1. Carlsson, J., Harms-Ringdahl, L. and Kjellen, U. 1983. *Industrial Robots and Accidents at Work*, TRITA-AOG0026. Occupational Accident Research Unit, Stockholm.
2. Tiefenbacher, F. 1982. *Industrial Safety and the Use of Industrial Robots*. Working Environment Fund, Stockholm.
3. Backstrom, T. and Harms-Ringdahl, L. 1984. A statistical study of control systems and accidents at work. *J. Occupational Accidents*, 6/84.

Appendix-Description of Accidents

Case 1

Robot: Manually controlled manipulator.

Industry: Manufacture of milking machines.

Occupation of injured person: Machine operator.

Course of events: While a lathe in the robot line was being monitored, the automatic operation mode was turned on. The worker was about to lift off a part from turning position when the robot came along to pick it up; the worker's hand was jammed.

Action taken: More information provided. Before parts are removed from their positions the automatic operation mode must be disengaged.

Injury: Bruise and pressure contusion on the right hand.

Sick leave: Probably more than 7 days.

Case 2

Robot: Manipulator, type 2P-DX1000.

Industry: Machine shop, manufacture of gears for cars.

Occupation of injured person: Electrical mechanic.

Course of events: When moving the robot, the operator happend to release the arm downwards via mechanical action of a valve. He got his hand caught between the damper and the robot arm's mechanical stopper.

Injury: Bruises on right hand.

Sick leave: Probably more than 7 days.

Case 3

Robot: Manually controlled manipulator.

Industry: Machine shop.

Occupation of injured person: Metalworker.

Course of events: Worker was busy marking door panels. The sheet metal used for this purpose is picked up from a conveyor belt by a robot. As the worker was removing tape from a sheet, another sheet was moving from the conveyor belt to the assembly line. He sustained a light pressure contusion on the middle finger of his right hand when the two sheets collided.

Injury: Light contusion on middle finger of right hand.

Sick leave: 4 days.

Case 4

Robot: Industrial robot.

Industry: Manufacture of automobiles.

Occupation of injured person: Repairman.

Course of events: Worker was busy carrying out overhaul work on the robot; he had changed the servo valves. In the course of a trial run the 'up-down' mode started moving because of air in the system and the repairman's left leg became jammed.

Injury: Neurotrasis in left leg, ligament wound.

Sick leave: Probably more than 7 days.

Case 5

Robot: Robot for automatic finishing device (made by Kimstaverken).

Industry: Metalworking plant, finishing department.

Occupation of injured person: Repairman.

Course of events: While copper rails were being mounted for degreasing vats, the robot was in operation. The repairman did not notice and he was squeezed against the vat and rails.

Action taken: Production will be halted when similar installation work is being done.

Injury: Cracked rib.

Sick leave: 12 days.

Case 6

Robot: Electrolux MHU, at injection moulder of Buhler make, type Revar.

Industry: Manufacture of kitchen ranges and refrigerators.

Occupation of injured person: Injection moulder.

Course of events: Worker was engaged in monitoring two injection moulders. One machine's conveyor belt had to be adjusted. While he was doing this he got in the way of the robot which works next to the moulder, and sustained a blow on the head from the robot.

Action taken: The robot must be turned off whenever somebody works inside its work area.

Injury: Head wound.

Sick leave: 4 days.

Case 7

Robot: AB Kaufeldt Pat: 2225-70.

Industry: Plastics plant, injection moulding department.

Occupation of injured person: Repairman.

Course of events: While doing adjustment work and removing oil from under the robot, the repairman's right foot slipped. He got in the way of the gripping arm since the machine had not been turned off.

Injury: Bruise resulting in slight brain concussion.

Sick leave: 3 days.

Case 8

Robot: Unimate.

Industry: Automotive industry, manufacture of gearboxes.

Occupation of injured person: Relief man.

Course of events: Worker was going to man a robot-served processing group. The robot, which serves a number of machines, is surrounded by a guardrail. Two gates with safety catches are coupled to the robot, and the robot is supposed to stop when passing through the gate. In the course of automatic operation the robot had shut down next to a delivery station. The worker opened the gate and went inside to determine the cause. In doing so, the robot suddenly started, jamming the relief man.

Action taken: The robot has been taken out of production. An investigation is in progress.

Injury: Bruise on left hand.

Sick leave: 9 days.

Case 9

Robot: Air-powered manipulator.

Industry: Metalworking plant, pressing department.

Occupation of injured person: Metalworker.

Course of events: An incorrectly inserted part got caught in the tool. The worker was going to remove it but had forgotten to evacuate the hose supplying air to the robot. When the part was removed, the robot continued its cycle, and the worker's hand got jammed between the robot and the tool.

Action taken: There are personnel at hand (set-up man, foreman) who are supposed to remedy such situations. The worker was provided with new information.

Injury: Crushing injury and pressure contusion of left hand.

Sick leave: 1 day.

Case 10

Robot: Manually controlled manipulator.

Industry: Manufacture of car radiators.

Occupation of injured person: Solderer.

Course of events: Worker was dip-soldering radiator cores with a robot to help him. The robot failed to grab the core, so using his right hand he tried to put the core in the right position. He got his middle finger jammed between the dipping pot and the robot's gripping rulers.

Injury: Bruise on middle finger of right hand.

Sick leave: 22 days.

Case 11

Robot: Manually controlled manipulator.

Industry: Plastics plant.

Occupation of injured person: Plastics worker.

Course of events: The robot had stopped because a part was not in the right position. The injured person was then going to adjust a part for correct positioning, whereupon he came in contact with a circuit breaker. As he did so, the robot assumed normal operation. It was when the robot was going to insert a new part into the tool that the injured person got jammed between robot and tool.

Injury: Crushing injury and pressure contusion of finger.

Sick leave: 2 days.

Case 12

Robot: ASEA.

Industry: Manufacture of alternating current machinery.

Occupation of injured person: Pressure moulder.

Course of events: Worker was repairing a machine in a pressure-moulding plant and turned round to get hold of a tool. A robot, which was then in operation, was placed behind. In turning round the worker struck the robot hand with his lower jaw.

Action taken: Admonition to run the robot arm in a position such that it cannot cause harm when machines in the vicinity are being repaired.

Injury: Four front teeth knocked loose.

Sick leave: None.

Case 13

Robot: Paint-spraying.

Industry: Automobile and aircraft plant.

Occupation of injured person: Robot operator.

Course of events: Operator was engaged in programming a robot to paint an intermediate coating. Steering the machine is rather heavy going. He twisted himself and sprained his back.

Injury: Back sprain.

Sick leave: 3 days.

Case 14

Robot: Electrolux MHU.

Industry: Manufacture of vacuum cleaners, manufacture of robots.

Occupation of injured person: Metalworker.

Course of events: The line for rotors to lathes from the robot was full. The worker then set out to remove four rotors. As he was doing so, the robot came along and placed a hot rotor on the worker's wrist.

Action taken: Marking of an emergency stop at the place where rotors are to be picked up.

Injury: Burn on right wrist.

Sick leave: Probably more than 7 days.

Case 15

Robot: Welding.

Industry: Manufacture of automobiles.

Occupation of injured person: Repairman.

Course of events: Repair of pliers. While giving the robot a test run the repairman happened to get his left index finger jammed owing to a misunderstanding between the operators.

Injury: Pressure contusion of left index finger.

Sick leave: 12 days.

Case 16

Robot: ASEA 1Rb60.

Industry: Manufacture of robots.

Occupation of injured person: Systems tester.

Course of events: A robot was being test-run at an enclosed testing place. Using his right hand, the worker was going to lean on (load) the robot while it was in motion. His right little finger got jammed between the robot arm and the screwed-on counter-weight.

Action taken: Personnel will be urged not to hold the robot while it is operating. Another person will stand 'at the ready' next to the emergency stop. The counterweight is to be marked so as to leave sufficient space for a hand.

Injury: Pressure contusion and bruise of right little finger.

Sick leave: 27 days.

Case 17

Robot: Manually controlled manipulator.

Industry: Machine shop, punching of sheet-metal rounds.

Occupation of injured person: Metalworker.

Course of events: Worker was to remove a round that had got stuck in the robot when the machine suddenly started. In jerking back his hand, he hit it against a round.

Action taken: Change made in safety instructions.

Injury: Bruise of right middle finger.

Sick leave: Probably more than 7 days.

Case 18

Robot: Manually controlled manipulator.

Industry: Manufacture of kitchen fans.

Occupation of injured person: Metalworker.

Course of events: Parts for circuit closers were being run in a manipulator. The worker was busy running parts in this robot. In order to pry loose a part that had got stuck, he opened an aperture and thrust his left arm before the machine had finished all operations. So the machine kept going and the worker's arm was jammed.

Injury: Light pressure contusion on the left arm.

Sick leave: 3 days.

Case 19

Robot: Unimate Robot Sv-13150.

Industry: Automotive plant, engineer shop.

Occupation of injured person: Operator at robot group.

Course of events: Alignment of traction machine. While attempts were being made to get this machine started for automatic operation, the robot was set in motion and jammed the operator's left lower arm.

Action taken: Change of clamping device at gate.

Injury: Pressure contusion on left lower arm.

Sick leave: 10 days.

Case 20

Robot: ASEA manipulator, 60 kg.

Industry: Manufacture of robots.

Occupation of injured person: Fitter.

Course of events: Worker was going to fetch some material. As he did so, he passed through a place for final assembly of a robot under construction. The upper arm of the robot was set horizontally, which he did not notice and walked right into the robot arm gashing his forehead.

Action taken: Information provided about the hazards involved.

Injury: Bruise of the forehead and probable brain concussion.

Sick leave: Probably 1 to 7 days.

Case 21

Robot: Picmat manipulator, used in conjunction with foundry machine.

Industry: Metal foundry.

Occupation of injured person: Set-up man.

Course of events: The injured person got in the way of the machine.

Injury: Bruised ear.

Sick leave: 3 days.

Case 22

Robot: No information available about type.

Industry: Manufacture of explosives and igniters.

Occupation of injured person: Machine setter.

Course of events: After adjustment of ELSA, a six-meter machine, the banderole on a coil winding had not been properly gathered together. As soon as the banderole was adjusted, which was done while the machine was in operation, the robot hit the upper side of the man's hand.

Action taken: The machine is fitted with protective railing. The setter has been informed about the hazards involved in making adjustments while the machine is turned on.

Injury: Bruise on right hand.

Sick leave: 4 days.

Case 23

Robot: Manual manipulator.

Industry: Manufacture of glass.

Occupation of injured person: Glassworker.

Course of events (as told by glassworker): "When you're working on the bigger glasses, the robot arm will let go of them and put them on a table for further cutting by hand. A piece of glass burst in the machine. I stepped back to see what was happening at the machine. I got the robot arm in my back. A workmate of mine saw the whole thing: he acted fast and pulled at the robot arm to make it let go."

Injury: Crushing injury and pressure contusion in the back.

Sick leave: 17 days.

Case 24

Robot: ASEA-60 manipulator used in conjunction with lathe Menforts SV-14452.

Industry: Automotive plant, manufacture of gearboxes.

Course of events: Rigging of robot. Robot got stuck in the lathe. Worker went to help out manually using his hand to pry loose the robot, which had come undone at the same time. The little finger of his right hand was jammed in the grab claws.

Action taken: Make sure your hands are not on the robot when it is being rigged.

Injury: Crushing injury and pressure contusion of little finger on right hand.

Sick leave: 17 days.

Case 25

Robot: Manually controlled manipulator.

Industry: Foundry and machine stop.

Occupation of injured person: Experimental mechanic.

Course of events: The injured person was jammed tight by the pick-up arm while doing adjustment work.

Injury: Contusion of finger.

Sick leave: 13 days.

Case 26

Robot: Type not registered.

Industry: Plastic manufacture, blow moulding.

Occupation of injured person: Plastics worker.

Course of events: While attending to the machine, the injured person was going to pick out a part. In doing so, he got in the way of the robot stationed alongside the machine. He sustained a light blow on his thumb.

Injury: Bruise of right thumb.

Sick leave: 10 days.

Case 27

Robot: Manually controlled manipulator used in conjunction with Velden automatic welder.

Industry: Welding of PVC articles.

Occupation of injured person: Plastics welder.

Course of events: A major accessory of the automatic welder that the person worked with is a robot that puts pasteboard into foil for in-welding. Here, the welder was going to make a manual correction without having to stop the machine. When the next batch of pasteboards arrived, he did not have time to move out of the way and was hit on the head by one of the robot's moving parts. As a result he was squeezed between the robot and a bar that is permanently attached to the machine.

Injury: Head wound, brain concussion.

Sick leave: 17 days.

Case 28

Robot: Manually controlled manipulator for pressing in a transfer plant, VVamac-Ida 6.

Industry: Engineering industry.

Occupation of injured person: Metalworker.

Course of events: Since the metalworker was going to start the robot, she had to connect an aerial line which the preceeding workteam had removed for cleaning. On reconnection, the robot went back to its given position and the worker sustained a blow inflicted by a protruding part at the rear of the robot.

Action taken: The aerial line is now disconnected in a safe spot. Protective devices are going to be installed around the robot.

Injury: Scalp wound and sprained neck.

Sick leave: More than 7 days.

Case 29

Robot: Manually controlled manipulator.

Industry: Manufacture of separators.

Occupation of injured person: Operator in automatic assembly line.

Course of events: In the course of rigging a robot to handle sheet metal in the manufacture of retainer plate, the operator lifted the sheet to adjust the position.

Injury: Sprained back.

Sick leave: 4 days.

Case 30

Robot: Lifting robot.

Industry: Manufacture of concrete roof tiles.

Occupation of injured person: Factory worker.

Course of events: The injured person stood in the way of the robot and was pushed in the back.

Injury: Backache.

Sick leave: Probably 1 to 7 days.

Case 31

Robot: Manually controlled manipulator.

Industry: Manufacture of electrical machinery.

Occupation of injured person: Tester.

Course of events: As part of a testing procedure for Zno blocks, the tester was going to polish the electrodes. He turned off the program and the input, but forgot that the fixture is controlled by its own program. The fixture descended and jammed his left index finger.

Action taken: The fixture will be rebuilt so that it cannot go into lowest position.

Injury: Contusion, fracture of left hand.

Sick leave: Probably more than 7 days.

Case 32

Robot: Material grabber KARMANN, special designs.

Industry: Manufacture of car cabins.

Occupation of injured person: Foreman.

Course of events: An electric sensor was to be reprogrammed. The machine worked as intended. The foreman wanted to help a repairman who was going to modify the function of a materials grabber. In carrying out testing procedures the foreman held onto the workpiece while the repairman slowly test-ran the grabber. As a result the foreman's fingers were jammed between the grabber's stopping device and the grabber itself.

Action taken: According to the instructions, no equipment is allowed to be operated in this manner. Point this out to the people concerned.

Injury: Pressure contusion on two fingers. One fingernail removed.

Sick leave: 2 days.

Case 33

Robot: Manually controlled manipulator.

Industry: Manufacture of electrical machinery.

Occupation of injured person: Tester.

Course of events: Worker was testing a Zno block when a fault developed. He turned off the robot program but forgot that the fixture is controlled by its own program. The fixture descended and jammed the middle finger of his left hand.

Action taken: The fixture will be rebuilt so that it cannot go into lowest position.

Sick leave: 10 days.

Case 34

Robot: Electrolux MHU junior.

Industry: Manufacture of electric components for power engineering purposes.

Occupation of injured person: Metalworker.

Course of events: Operation of automatic metal plater. When trouble-shooting the device's punching unit, the operator would run the robot manually. The operator wrongly manoeuvered the robot's arm, making it get stuck in the metal hopper's vertical picking cylinder. In so doing he triggered off the robot's emergency stop. The device's pneumatic part will not be evacuated when the emergency stop is activated. The operator was informed of this. He then loosened a barring chain and entered the robot's work area to loosen the robot arm from the picking cylinders. Due to the accumulation of compressed air, the robot moved at great speed in its return mode, jamming the worker's thumb against the terminal stop. The compressed-air vent had to be shut off and the air evacuated before the barring could be interrupted.

Action taken: Coupling of electric turn off/evacuation valve to emergency stop.

Injury: Pressure contusion on left thumb.

Sick leave: Probably more than 7 days.

Case 35

Robot: Coil-winding robot MICAFIL.

Industry: Manufacture of electrical machinery.

Occupation of injured person: Instructor.

Course of events: Winding of laminated core in a coil-winding robot. The machine worked as intended. A protective hood is fitted around the machine's winding head. The instructor found himself in the protective hood so as to change the strip rollers. In order to get one strip holder into the desired position, he stretched out to reach the drive control outside the hood and started the machine. The other strip holder then seized his left thigh and pressed his foot against the floor. Normally, nobody is supposed to be inside the hood while the machine is on.

Action taken: Mounting of readily visible emergency stop on outside of hood.

Injury: Bruise of left thigh and sprained ankle.

Sick leave: 5 days.

Case 36

Robot: Fixed-sequence robot (for assembly).

Industry: Aluminium foundry.

Occupation of injured person: Furnace mason.

Course of events: In the course of working on the furnace, the mason moved the robot's metal scoop from the furnace to a neutral position. By mistake the scoop came to a halt halfway. The mason went underneath to look and leaking metal dripped on his neck.

Injury: Burn in neck.

Sick leave: 9 days.

People and Robots – Their Safety and Reliablity

R. Jones and S. Dawson
Centre for Robotics and Automated Systems,
Imperial College of Science and Technology, UK

Data on the performance of 37 robot systems in three companies covering 20,000 robot production hours are presented. The research aims and method of the data collection and analysis are outlined. The bulk of the paper is devoted to the presentation of preliminary findings concerning the reliability and safety of the installations. The main impression gained so far is that robot installations do give some ground for concern about variable reliability and potential for injury and harm. Some wider implications are mentioned, including particular reliability problems and the importance of training in mitigating possible hazards.

The aim of the project is to collect performance data on robot use to identify problems and assess the implications of these problems for safety and reliability. Strategies and tactics which management and specialist personnel may adopt to deal with these problems in terms of the design, implementation and use of robot systems are also considered.

During the second half of 1983, data was collected from six companies, about 84 robots covering five robot tasks. Table 1 shows the distribution of robots between companies and tasks. There was considerable variation between companies as regards their experience of robot use. Some had more than six years experience whereas others were just starting to introduce robots into their production processes. The range of tasks and robot types covered in the study is wide enough to be representative of a large proportion of the robots in use in the UK.

Data collection

The aim was to collect data on all incidents that occurred to disturb the normal course of operations of each robot installation between two and six months. The length of time for data collection varied according to the starting date and other commitments in the plant. The original intention was to ask the people directly responsible for each robot installation to record data on each incident as it occurred. This was done in Companies D, F and H where the person responsible for the robot completed a diary form supplied by the researchers for each incident. In Companies A and B, in-house records were already being collected on the performance of the robots and this information was used directly. It did not prove possible to set up a comparable data collection system in Company C because workloads on those responsible for the robots were already very high, and time could not be spared for the extra form-filling required. However, other useful data were collected from this company although they are not directly comparable with the data from other companies.

Data from Companies D, F and H were collected and analysed, and data from Companies A and B were analysed, in terms of the following categories:

- Incident date.
- Robot's number or other identification.
- Downtime occasioned by the incident.
- Reasons for action being taken.
- Actions taken.
- Personnel involved in identifying or noticing the problem and in sorting out the problem.
- Details on the incident, including the means of interruption and classification of incident according to the consequence.
- The recorder's assessment of the underlying reason for the problem.

Fig. 1 shows the ways in which data in some categories were analysed.

Analysis of the underlying reason for the incident proved difficult. To begin with, a variety of factors are likely to be important and some of them are likely to be more easily apparent than others. Furthermore there was considerable variation in the level of detail given, which also tended to be highly specific to the robot installation being considered. Fig. 2 shows the list of categories which have been used to describe the underlying causes. They have the merit of enabling comparative analysis between companies and between robot tasks. Faults and problems directly related to the robot installation are analysed in greater detail than other 'underlying reasons' since they are of particular interest in this study.

Data generated from diaries of performance in Companies D, F and H and the in-house records of Companies A and B were complemented by substantial observation of the manufacturing process and the practices involved in each case. Discussions were also carried out between the researcher and those conversant with the robots in each company, including foremen, engineers, operators, safety officers and safety representatives. The preliminary findings are divided into two sections. The first deals with the reliability of the robot installations and the second with the safety aspects of their use.

At this comparatively early stage of the project, comparable data is available on over 20,000 robot hours, distributed between 37 robots of four different designs in three companies (A, B and F). The appendix provides a glossary of created variables used in this analysis showing their definitions and methods of calculation. It should be used as a reference point for terms used in Tables 1–8.

Reason for action being taken	Action taken	Personnel involved
1. Mechanical problems (a) robot (b) other machinery	1. Replacement of faulty equipment	1. Maintenance fitter
2. Electrical problems (a) robot (b) other machinery (c) interface	2. Adjustment to equipment/cleaning	2. Electrician
3. Inspection of process	3. Resetting equipment	3. Production engineer
4. Programme problem	4. Reprogramming	4. Operator
5. Quality problem	5. Routine (preventative/planned maintenance)	5. Foreman
6. Erratic robot behaviour	6. Unplanned maintenance (following an incident)	6. Chargehand
7. Dropped part	7. Fault diagnosis	7. Leading hand
8. Threatened damage (a) robot (b) persons (c) machinery	8. Other	8. Service engineer
9. Other		9. Other

Means of interruption	Classification of incident
1 Personal action (a) emergency stop (b) other personal action	1. Accident (with actual harm) (a) person (b) machine
2. Automatic action (a) controlled stop (b) signals from sensory equipment	2. Accident (with actual damage) (a) person (b) machine
	3. Near miss (damage or harm narrowly missed)
	4. Incident (but no harm or damage occurred)
	5. Hazard anticipated/preventative action
	6. No damage or harm likely to occur

*Where more than one reason, action or person involved, more than one code was recorded for each incident

Fig. 1   Data collection and analysis for each incident

1. Robot related problems

Category	Definition
Component failure in robot arm	A discrete, known failure in the robot arm
Fuses blown	The fuses in the robot's controls fail for various reasons–including faults elsewhere
Fault in cabinet	A recognised fault in the components of the control cabinet, e.g. a faulty circuit board
Fault in teach pendant	Component failure in teach pendant, e.g. a faulty button
Power supply fault	The power supply to the robot fails (particularly with hydraulically powered robots–possibly due to a faulty filter or dirty fluid causing a blockage)
Cable/transmission problem	Broken cables or other similar transmission difficulty in robot unit
Overheating hydraulics	Overheating of hydraulic power pack causes a failure of power and freezes the robot's motion
Robot collision	Robot collides with other equipment or another robot
Robot won't move	Robot fails to move, though no apparent reason is found–possibly a software problem

2. Problems not directly attributable to robot units

Category	Definition
Other component failure	Component failure on equipment other than the robot
Other equipment problem	Production equipment other than the robots have caused a problem (but not a failure of a component)
Sequence fault	The normal sequence of events is disrupted by something unspecified–possibly transmission
System failure	The whole system fails to work with no single identifiable component at fault
System checks	Checks are carried out on the system as a means of preventing problems becoming worse
Check on parts	Checks are made on components as they progress through the robot system
Part problem or variation	The components are at fault, e.g. because of excessive variation from design parameters
Quality problem	The quality of the process has been identified as sub-standard
Services problem	Problem with services, e.g. electricity, water, compressed air, other gas supplies
Safety function problem	Problem identified with safety equipment, e.g. interlocked gates, fencing, light guards

Category	Definition
Robot out of synchronisation	Robot arm position does not match the robot's memory of where it should be—commonly referred to as 'out-of-position' or 'lost itself' or 'lost wait position'
Robot in emergency stop	The controls bring the arm to a halt—referred to as 'stop-in-sequence', 'trip-out' or 'lost control on robot'
Erratic robot	The robot moves off the specified path by an appreciable amount
Stiffness in robot	Mechanical problems cause the arm to resist motion—results in the controls stopping the motion
Problems in tools	A fault is found in the tools carried on the end of the robot arm such as weld guns and their cables
Robot problems— no detail	Any robot related problem not covered above

Category	Definition
Human error	Incorrect human action resulting directly in a loss of production
*Weld failure	The robot failed to carry out it's welding tasks; this could be due to a number of things
Process problem	The process has been shown to be faulty in some other way than above

*In Company A, the robot systems and data collection were such that it was often impossible to specify the underlying reason for a fault in any greater detail than 'weld failure'. It will be seen that such cases accounted for more than 50% of incidents from this company. The robots' tasks were arc welding and the first indication of a fault was an alarm signalling the failure of the robot to carry out it's task. In each case, a number of factors could have been important in failure. For example, part variation or some problem with the weld equipment or some unspecified erratic robot motion. In many cases it was considered that in fact the problem was robot related and should therefore have been ideally categorised in Section 1.

Fig. 2   Classification of underlying reasons for incidents or problems

Table 1   Robots and their tasks in the companies studied

Company	Arc welding	Spot welding	Materials handling	Adhesive bonding	Routing	Total
A	12	–	–	–	–	12
B	–	52	–	2	–	54
C	7	–	–	–	–	7
D	–	–	2	–	–	2
F	5	2	–	–	–	7
H	–	–	–	–	2	2
Total	24	54	2	2	2	84

Reliability of robot installations

In each of Companies A, B and F the overall downtime figures were surprisingly high. Overall downtime accounted for more than 22% of production time in each. Taking all incidents in these three companies together, 23.4% of the cases and 21.4% of the downtime were directly connected with robot problems (see Tables 2 and 3).

Major problem areas

Table 2 shows the overall picture of downtime in each company. One could be drawn to the conclusion that there is a great deal of similarity between the companies from the way the total amount of downtime is approximately a quarter of production time in each case. However variations between the companies and robot tasks become apparent as soon as one looks at the underlying reasons for the incidents. Table 4 gives details on the main reasons which emerged from each company. The percentage of cases attributable to robot related problems is given for each company. Other reasons are included if they account for either 10% or more of cases, or 10% or more of the downtime.

In Company A, weld failures account for a majority of cases but less than 10% of the total downtime. This is in part due to the generally short duration of each incident. Moreover, with a high number of cases of weld failure (58.6%)

Table 2   Overall downtime figures*

Company	Robots studied	System production hours	Incidents recorded	No. of incidents with no downtime	No. of incidents with downtime	Percentage of production time	System downtime (minutes)
A	12	866	743	357	386	23.5%	12223
B	23	396	616	81	535	22.6%	5366
F	2	1216	811	6	805	26.7%	19486
Overall	37	2478	2170	444	1726	24.9%	37075

*Robot production hours covered: 21,932
See Appendix for definition and calculation of terms

Table 3   Robot related problems, RRP*

Company	No. of cases of RRP	Percentage of all cases	Downtime for RRP (minutes)	Percentage of downtime for RRP	No. of cases with no downtime	Robot related downtime as percentage of production time
A	47	6.4%	949	7.8%	27	1.8%
B	306	49.6%	3248	60.5%	35	13.6%
F	155	19.1%	3737	19.2%	1	5.1%
Overall	508	23.4%	7934	21.4%	63	5.3%

*As defined in Fig. 2
See Appendix for definition and calculation of terms

giving no downtime the proportion of the total recorded downtime is artificially low. Company A has a far higher proportion of cases with no downtime figures than the other two companies. Two main reasons account for this. One reason relates to the layout of the robot system, the other to the difficulties of keeping records. For certain incidents, like weld failures, only part of the system may need to stop. This could lead to the rest of the system stopping if the incident endured for long enough. It was difficult to work out the system downtime caused on such occasions. Besides this, for incidents of short duration it was found that the time was not always meticulously recorded. Average downtime figures for cases in Company A were generally higher than for the other two companies. Excluding weld failures, all the average downtime figures for Company A shown in Table 4 were over 45 minutes, whereas for Companies B and F, only component failure cases were above this.

In Company B, robot related problems account for over 60% of the total downtime as well as nearly 50% of cases. This is far more than in the other two companies. Other equipment problems were of less importance in Company B. For Company F 'other equipment problems' and component failures together accounted for over 60% of the downtime and nearly 50% of cases. These variations are in part due to the different layouts and equipment employed in each installation. Some of these variations as far as robot related problems are concerned are discussed below.

Robot related problems
One can see from Tables 3 and 4 that robot related problems in Company B constitute a far higher proportion of cases and of downtime than in the other two companies. However, it should be noted (see Table 2) that Company B has the largest number of robots as well. This point is not taken into account by any comparisons made from Tables 3 or 4. One needs to look at robot related problems and downtime in conjunction with robot production hours. The downtime as a percentage of robot hours can then be expressed, that is, the amount of time on average that each robot is down. An approximation of the mean time to a robot related problem for each robot can also be produced (see Table 5).

Table 4   Major problem areas with robot installations in each company*

Problem	Percentage of cases	Downtime (minutes)	Percentage of total downtime	No. of cases with no downtime	Average downtime (minutes)
Company A					
Robot related problems	6.4%	949	7.8%	27	47.45
Other equipment problems	12.9%	3694	30.2%	39	64.8
Sequence faults	6.1%	2062	16.9%	14	66.5
Quality problems	5.5%	1845	15.1%	22	97.1
Weld failure	51.3%	1113	9.1%	209	6.5
Total	82.2%	9663	79.1%	311	32.3
Overall	100%	12223	100%	357	31.7
Company B					
Robot related problems	49.6%	3248	60.5%	35	12.0
Other component failure	2.3%	591	11.0%	3	53.7
Other equipment problems	27.3%	534	10.0%	16	3.5
Total	79.2%	4373	81.5%	54	10.1
Overall	100%	5366	100%	81	10.0
Company F					
Robot related problems	19%	3737	19.2%	1	24.3
Component failures	16.6%	6253	32.1%	—	46.3
Other equipment problems	31.1%	5569	28.6%	3	22.4
Quality problems	23.1%	2477	12.7%	2	13.4
Total	90%	18036	92.6%	6	24.7
Overall	100%	19486	100%	6	24.2

*Underlying reason as defined in Fig. 2
See Appendix for definition and calculation of terms

Table 5 shows that in fact Company F appears to have the worst robot record rather than Company B. Company F has a mean time to robot related problems of 15.3 hours and a downtime figure for robot related problems of 2.6% of robot hours. Both these figures are significantly worse than for the other two companies, but even then they suggest a fair degree of reliability for the robots themselves. That Company A's figures are much better than those of the other two companies, must be in part due to the high frequency of weld failures which, it will be recalled, are not included as robot problems in this analysis (see footnote to Fig. 2). And yet, it is felt they often have robot related problems as their cause.

Table 5   Robot related downtime expressed as proportion of robot hours

Company	Robot production hours	Robot related downtime as percentage of production time	Robot related downtime as percentage of robot hours	Mean time to a robot related problem (hours)
A	10392	1.8%	0.15%	220.8
B	9108	13.6%	0.6%	29.6
F	2432	5.1%	2.6%	15.3
Total	21932	5.3%	0.6%	42.9

See Appendix for definition and calculation of terms

Table 6 Detailed examination of robot related problems

Problem	Company A		Company B		Company F	
	%company downtime	average downtime (mins)	%company downtime	average downtime (mins)	%company downtime	average downtime (mins)
All robot related problems	7.8%	47.45	60.5%	12.0	19.2%	24.3
Component failure in robot arm	1.5%	180.0	2.8%	21.4	3.3%	216.6
Fuses blown	–	–	3.4%	16.7	–	–
Fault in cabinet	3.4%	420.0	3.6%	65.0	1.8%	116.7
Fault in teach pendant	0.1%	10.0	–	–	–	–
Power supply fault	–	–	–	–	2.4%	9.0
Cable/transmissions problem	–	–	3.3%	19.6	2.2%	52.5
Overheating hydraulics	–	–	–	–	0.7%	12.7
Robot collision	–	–	5.2%	11.7	–	–
Robot won't move	0.4%	26.0	1.9%	16.8	0.05%	10.0
Robout out of synchronisation	0.1%	13.0	2.7%	6.0	0.5%	47.5
Robot in E stop	0.07%	2.7	5.9%	2.9	0.3%	8.6
Erratic robot	1.3%	77.5	9.0%	40.2	2.9%	13.1
Stiffness in robot arm	–	–	0.4%	12.0	0.05%	10.0
Problems in tool	0.7%	17.6	19.4%	17.7	1.5%	15.8
Robot problem— no detail	0.2%	5.75	2.9%	22.0	3.6%	115.7

See Appendix for definition and calculation of terms

Turning to look at the robot related problems in some detail, Table 6 presents a comparative analysis of the underlying reasons, showing the percentage of the total downtime involved and the average downtime caused. In certain cases, the average downtime seems to be very high in comparison with the percentages. This is especially so in Company A with 'component failure in robot arm', 'fault in robot cabinet'. This was in fact due to the downtime resulting from one case in each category. The contribution to downtime should therefore probably be regarded as more important than the contribution to the number of cases. This point is also brought out by Table 7 which examines actual robot failure categories, i.e. cable problem, component failure in arm, blown fuses, fault in cabinet, teach pendant fault and power supply fault. This shows that though the contribution to robot related

Table 7 Actual robot failures

Company	Cases of robot failure	Percentage of all cases	Percentage of robot related downtime	Percentage of total downtime	Mean time to robot failure (hours)
A	4	0.5%	64.1%	5.0%	2595.5
B	32	5.2%	21.7%	13.1%	284.3
F	65	8.0%	50.5%	9.7%	36.9
Overall	101	4.7%	40.2%	8.6%	216.6

See Appendix for definition and calculation of terms

Table 8   Cases with short and long periods of downtime

	Cases of downtime of less than 10 minutes			Cases with downtime of more than 1 hour		
	Company A	Company B	Company F	Company A	Company B	Company F
Cases	262	440	438	37	19	60
% of all cases	35.3%	71.4%	53.8%	5.0%	3.1%	7.4%
Downtime (mins)	1231	1539	3389	8785	2056	9029
% of total downtime	10%	28.7%	17.4%	72%	38.3%	46.3%
Average downtime (mins)	4.7	3.5	7.7	237.4	108.2	150.5

See Appendix for definition and calculation of terms

downtime is quite significant, the number of cases involved are few. One might consider that with such a low frequency of occurrence they were not major problems but the downtime caused is always proportionately greater. This point is also brought out by an examination of the relevant average downtime figures in Table 6.

The quantitative data suggested that erratic or unspecified robot motion is a fairly frequent occurrence. In Companies A, B and F it was taken as underlying cause of between 1% and 5% of cases. Though the erratic motions may be only 20mm off the programmed path, they can result in potentially serious incidents, for example the robot colliding with a fixed object, such as a jig. Damage to the robot or the tool carried could occur as a result. Several cases have been noted in this study causing considerable loss of production and repair costs. Sequence faults have also caused damage to a few robots in the study.

A comparison of cases with short and long periods of downtime
Considerations of cases with short and long periods of downtime facilitates a comparison between cases which occurred frequently but with short periods of downtime, and those which are less frequent but are associated with long periods of downtime. Table 8 shows cases with periods of downtime of 10 minutes and less, and those of 1 hour or more. In Company B the percentage of downtime caused by short incidents is fairly close to that for the long incidents. In the other two companies, and particularly Company A, incidents with long periods of downtime dominate the total. However, frequent incidents of short duration are an important feature in all three companies.

Safety aspects of robot use

People at risk
Previous work in the area has been carried out predominantly in countries with a higher concentration of robots than the UK. Research in Sweden[1] (see also page 49) and Japan[2] has suggested incidents where

accidents and near accidents are likely to occur in robot installations. Particularly hazardous periods were shown to occur during close man–machine contact, when people are involved in such activities as reprogramming, repairs, making adjustments or cleaning. Since such actions require the robot system to be non-productive for a period, they have been covered by this study. However this does not mean that all incursions into robot installations were recorded since observation showed that in certain cases access was possible with no loss of production and no downtime being caused. One should not assume that hazards are not present during such times. Working practices during periods of close man–machine contact where the system remained productive were observed by the researcher, even though they were not recorded in the quantitative data.

Some serious problems did occur during the field work period of this study, although they were infrequent. No cases of injury resulting from direct contact between a person and a 'rogue' robot occurred in any of installations whilst the data was being collected. One injury was recorded, but this was due to problems with equipment rather than the robot. The people who are going to be most at risk in robot installations are clearly those who come into direct contact with them. The groups of employees who are involved vary with the working practices, between installations. In one of our cases, the running of the robots was given over exclusively to maintenance workers and their foreman. In another the production chargehand on each shift along with their leading hands were expected to sort out all the problems as they occurred only calling in the maintenance function for tasks which proved too difficult for them. Generally then from a safety viewpoint, fitters, electricians, foremen and chargehands are exposed to the greatest risks. It was these groups which, as Table 9 shows, predominated amongst those who were involved in identifying and sorting out problems. The data shows that operators rarely become involved in sorting out a problem although observation suggested they did play a more prominent role in cases of minor significance which went unrecorded.

Remedial action

It will be recalled that data was collected on the action taken in respect of each incident. Table 10 shows the results of a preliminary analysis of action taken in Companies A, B and F.

Table 9   Involvement of people in recorded incidents

Group of employees	Identification of problem (% of cases)	Sorting out problem (% of cases)
Fitter	34.1 %	51.5 %
Electrician	59.7 %	64.8 %
Production engineer	0.15%	0.07%
Operator	3.3 %	0.1 %
Foreman	26.5 %	26.1 %
Chargehand	19.9 %	28.0 %
Leading hand	5.5 %	16.8 %
Service engineer	0.03%	0.4 %
Other	4.2 %	7.2 %

Table 10   Action taken in respect of each recorded incident

| Action taken | Percentage of cases where each action is recorded | | | |
	Overall	A	B	F
Replacement of faulty equipment	15.2%	3.6%	19.0%	22.8%
Adjustment/cleaning	25.9%	7.0%	16.4%	50.3%
Resetting	58.3%	72.4%	64.0%	41.2%
Reprogramming	8.1%	8.1%	8.9%	7.6%
Routine maintenance	10.6%	2.0%	5.7%	22.3%
Unplanned maintenance	33.2%	67.4%	21.0%	11.1%
Fault diagnosis	3.0%	5.9%	2.3%	1.0%
Other	5.1%	6.6%	1.6%	6.3%

To understand the implications of these figures for safety one must consider the operating state of the robot systems during any action taken. For example, in each company studied, resetting in fact involved minimal contact with the system and could be carried out from outside the safeguards. However, the system had to be 'live' in nearly all companies, if reprogramming, fault diagnosis and to some extent, unplanned maintenance were to be carried out. Clearly the risks associated with an employee being in close proximity to a 'live' system are considerable. Data on actions taken cannot be linked directly with downtime figures because several separate actions can take place during any one period of downtime. For example, imagine that both resetting and reprogramming are carried out to rectify the fault of a robot being off its programmed path. It is for this reason that the columns in Table 10 do not add up to 100%. A full discussion of exposure to risk in robot installation is hampered by the early stage of the analysis.

Classification of the incidents
One can draw conclusions from the data on the classification of incidents for the severity of the problems. Only one incident was noted in the downtime records where an accident to a person occurred but this was unconnected with the robot itself. Table 11 however shows there were quite a number of near misses, accidents without damage to machinery or people and accidents involving damage to machinery.

Table 11   Classification of incidents

| Classification | Overall | | Company A | | Company B | | Company F | |
	No.	%	No.	%	No.	%	No.	%
Accident to person	1	0.0%	0	0.0%	0	0.0%	1	0.1%
Accident/damage to machinery	57	2.6%	10	1.3%	30	4.9%	17	2.1%
Accident no damage	15	0.7%	5	0.7%	7	1.1%	3	0.4%
Near miss	3	0.1%	1	0.1%	0	0.0%	2	0.2%
Incident no damage	1143	52.7%	554	74.6%	450	73.1%	139	17.1%
Hazard anticipated	13	0.6%	0	0.0%	0	0.0%	13	1.6%
No damage likely	908	41.8%	173	23.3%	126	20.5%	609	75.1%
Not given	30	1.4%	0	0.0%	3	0.5%	27	3.3%
Total	2170	100%	743	100%	616	100%	811	100%

Table 12 Downtime caused by accidents causing damage to machinery

Company	Cases	Cases with downtime	Downtime (mins)	Average downtime (mins)	Percentage of total downtime
A	10	6	768	128.0	6.3%
B	30	28	1230	43.9	22.9%
F	17	17	1085	63.8	5.6%
Overall	57	51	3083	60.5	8.3%

The table presents the number and frequency of each classification overall and in each company. One can see that the vast majority of incidents are of cases where no accident occurred or was likely to occur. In the few cases where accidents or near accidents were recorded, they effected damage to machinery rather than to people.

Another measure of the severity of these cases is expressed in terms of downtime which is shown in Table 12 for cases involving damage to machinery. It can be seen that such accidents account for an appreciable amount of downtime in each of the companies and particularly so in the case of Company B. The average downtime figures in Table 12 are also high in comparison with overall averages of downtime as shown in Table 4. Accidents causing damage to machinery are obviously serious from the production point of view. Their safety implications are also important. Since employees are usually required to repair the damage they are likely to be exposed to the possibility of a repetition of the incident. This is particularly true in the frequently undertaken exercise of checking the running of the machinery immediately following repair.

Means of interruption
Observation showed that under normal conditions workers did not resort to the use of the emergency-stop facilities in the robot systems studied, for spurious reasons. The frequency of their use can then be taken as a measure of the severity of the hazards encountered or anticipated by workers. Out of a total of 2170 cases, their use was recorded 11 times. This frequency of 0.005 overall matches the frequency of incidents being classified as 'hazard anticipated'.

Some implications for reliability and safety

Many of the implications of the findings of this study for reliability and safety will be the subject of future publications, once analysis of data from all companies is complete. For now we have chosen to limit ourselves to some comments on reliability and a brief discussion of the implications of the findings for training.

Reliability
Preliminary findings suggest there are a number of reliability problems for robot systems even if not to the same extent for the robots themselves. Certain specific points can be mentioned at this stage:

- *Power supply faults in Company F.* Hydraulically powered robots were used in this company. Of the cases 6.3% were found to be due to the power unit failing and a further 1.4% due to overheating in the hydraulics unit. Table 6 shows that 3.1% of the total downtime resulted from these problems. Company F was the only one reported here that used hydraulic robots. This type of failure largely accounts for the poor mean time to a robot failure shown in Table 7 for Company F.

- *Arc welding in Company A.* Arc welding appears to be very demanding for the present range of robots. This can be seen by the high proportion of the total cases (51.3%) with weld failures as the cause in Company A (see Table 4). However it was not the major cause of downtime even in this company.

Availability

Availability – the proportion of useful production time machinery is available for work – was not particularly high in any company. The overall downtime figures presented in Table 2 show that the robot systems were only available for about 75% of production time. But Table 3 shows that only a fifth of overall downtime was due to robot related problems. Robot availability in each system varied between 86.4% and 98.2%, with an overall figure of 94.7% (see final column of Table 3). Robots appear all the more reliable once the number of robots in each system is taken into account, as in Table 5. All this suggests that it is the systems as a whole and not the robots alone which are the main cause of 'unavailability'. The demands with which robot systems operate place production on a precarious path, with little variation in parts, tolerances or accuracy being accommodated.

Role of in-house maintenance

Generally the maintenance personnel within plants were capable of handling normal problems encountered in production with robots, but severe and unexpected problems and failures with the robots were not so easily dealt with. Many of the incidents of actual robot failures proved to be beyond the limits of the present expertise of in-house maintenance. This is perhaps not surprising since in the normal run of events the problems are not common. When such rare events occur, however, the recourse to suppliers' service engineers can have a considerable effect upon production. For example, in one case where a faulty axis board caused 'twitching' of the robot arm, most of the month the robot was out of service was spent by in-house electronics specialists diagnosing and attempting to rectify the faulty component. They successfully diagnosed the problem but did not succeed in rectifying the fault. When the service engineer was called he corrected the problem in one day. This company had expertise in electronics well above that of the others studied, but even so suffered because of their unfamiliarity with this type of production equipment. In other examples elsewhere the service engineer had to diagnose the problem to begin with and then sometimes return later with the correct part.

The diary data collected in Companies A, B and F shows however that the presence of a suppliers' service engineer is comparatively rare. None of these companies had the service engineer present for as much as 0.5% of all recorded cases. In general the length of experience with robot use was found to correlate with the ease of maintenance. Company B had the longest experience of robot use of the three companies dealt with in this paper. As Table 4 shows, Company B's average downtime figures were better than the other two companies.

Training for safety and reliability

The preliminary results contain several points of interest for the training of personnel. Some severe problems have been reported and hazards discussed. Though these may be rare, they have considerable importance in terms of the downtime produced and the risks posed. There is a need for training to point out the types of hazards that occur to those who are likely to become involved. The observation and discussion which accompanied the collection of quantitative data showed that when new robots are introduced there is a tendency for relatively senior people to be involved in the early stages. These people may be both specifically trained and fairly conversant with recommended safe working practices, particularly if they have just returned from suppliers' courses. When the installations become better established a number of practices may develop which have safety implications. First, the information gained from suppliers may become less clear with time and not fully shared with those who are to take over operating and maintaining the robot systems once they are in routine use. The people who assume responsibility for the robots may actually receive a lower level of training than their predecessors and thus not be as fully aware of possible hazards. The need for refresher courses is also an issue which should be borne in mind. Secondly, formal operating practices may prove to be impossible to implement fully whilst achieving production schedules. Even if this is not the case practices may deteriorate over time as familiarity takes the place of due respect for the capabilities of the system. A suggested increased frequency of entering the enclosure with lower levels of safeguarding than before may be symptomatic of the effects of familiarity.

Management have a clear role in maintaining safe and workable practices and ensuring that bad practices do not gain a foothold. This requires frequent monitoring to keep awareness of safe practices high[3]. Inspection by someone with knowledge and authority on the safe operation of robot systems would play a useful part in this. There is clearly a role for training safety officers and inspectors for this purpose.

Concluding remarks

This paper provides the first installment of results from our project to investigate reliability and safety aspects of robot installations. It has aimed to demonstrate that a comprehensive collection of performance data on robot installations is feasible and worthwhile. It has illustrated some of the ways in which the data can be analysed and some of the uses to which the analysis can be put. Further detailed analysis will we hope reveal increasingly interesting and useful findings. In particular, an Event Tree Analysis of selected problems or failures in a number of typical robot installations will be undertaken. This technique will describe the consequences of selected incidents for the operating state of the robot system considered, taking into account such things as the presence or absence and method of operation of a number of safeguards. For example, two initiating events worthy of consideration would be a sudden erratic motion of the robot or a sequence fault putting the robot out of phase with other equipment. Such an analysis will identify specific high risk occurrences and be capable of showing what safeguards are likely to contain the hazards produced.

The main impression gained from the project at this early stage is that robot installations do give some grounds for concern about variable reliability and potential for injury or harm. The managers and supervisors who discussed their robot installations with the researcher, often prefaced their remarks with words to the effect that "we don't have much trouble with our robots". The data presented here suggest a rather different view. Even though the robots themselves are fairly reliable, there are problems with the overall systems. Approximately 25% of system production time was lost in each of three companies, due to incidents which disturbed the normal running of the robot installations. Some of these incidents not only lost valuable production time, but they also placed people and production equipment at risk.

References

1. Carlsson, J., Harms-Ringdahl, L. and Kjellen, U. 1980. *Industrial Robots and Accidents at Work*. Occupational Accident Research Unit, Sweden.
2. Research Institute of Industrial Safety, Ministry of Labour. 1982/83. *Safety Measures of Industrial Robots*, Vols. 1 and 2.
3. Dawson, S., Poynter, P. and Stevens, D. 1983. How to secure an effective health and safety programme at work. *OMEGA, The International Journal of Management Science*, 11 (5): 433–446.

Appendix – Glossary of created variables

Cases

Percentage of cases. This is the number of cases under consideration expressed as a percentage of the total number of cases for each company:

as in Tables 3 – of robot related problems,
            4 – of major problem areas,
            6 – of robot problems in detail,
            7 – of actual robot failures,
            8 – of cases with short and long periods of downtime,
            9 – of employees involved in identifying or sorting out problems,
          10 – of remedial actions taken,
          11 – of the classification of incidents.

Downtime

Downtime totals. These are the sums of the downtime values of the cases of the category being considered:

as in Tables 2 – of total system downtime,
         3 – of robot related problems,
         4 – of major problem areas in each company,
         8 – of cases with short and long periods of downtime,
        12 – of damage to machinery cases.

Percentage downtime. This is the amount of downtime shown in a table expressed as a percentage of a specified total:

as in Tables 2, 3, 4 – as a percentage of system production hours,
        3, 4, 5, 6 – as a percentage of total downtime for each company,
            5 – as a percentage of robot production hours,
            7 – as a percentage of robot related downtime.

Average downtime. This is the total downtime given for the category under consideration divided by the number of cases with downtime figures:

as in Tables 4 – for major problem areas in each company,
         6 – for robot related problems in each company,
         8 – for cases with short and long periods of downtime,
        12 – for cases involving damage to machinery.

Mean time

Mean time to a robot failure. This is the average amount of productive robot time before a robot failure in each company. It is calculated as below:

$$\text{MTTRF} = \frac{\text{robot production hours} - \text{downtime due to robot failure (in hours)}}{\text{number of cases of robot failure}}$$

as in Table 7.

Mean time to a robot problem. This is the average amount of productive robot time before a robot related problem arises (in any company). This is calculated as below:

$$\text{MTTRP} = \frac{\text{robot production hours} - \text{downtime due to robot related problem (in hours)}}{\text{number of cases of robot related problems}}$$

as in Table 5.

Robot production hours

This is calculated as the product of the system production hours in each company and the number of robots studied in that company (as in Tables 2 and 5).

System production hours

The total number of hours of production time covered by the data collection period in each company. Note, this includes the total system downtime (as in Table 2).

Fault-Tree Analysis of Hazards Created by Robots

N. Sugimoto and K. Kawaguchi
The Research Institute of Industrial Safety, Ministry of Labour, Japan

From the standpoint of industrial safety, industrial robots present opposing characteristics. On the positive side, the introduction of industrial robots into dangerous and harmful operations once performed by humans has increased safety performance. On the negative side, however, industrial robots themselves can create dangerous conditions. To help ensure that the dangers inherent in industrial robots may not offset their usefulness, an attempt is made here to analyse available data on accidents and near accidents. On-site surveys of robot work to point out dangerous operations involving industrial robots are also analysed to compile a document on fundamental technologies and appropriate safety measures.

Although safety measures for robots can be established by observing existing regulations on labour safety and health, it is also true that dangerous operations exist in which none of these safety and health regulations have been applied. This paper attempts to identify and analyse them and their causes. It is impossible, however, to determine inherent dangers from the definitions of industrial robots available. Thus, the danger of industrial robots is discussed in relation to these robot characteristics: those with a free arm (including those with a grasping device at the arm tip), those with powerful arms, and those which are self-controlled.

As available accident data demonstrates, the possibility exists that the arms of these machines may unexpectedly go out of control, posing a danger to the operator. Such a threat applies to the majority of industrial robots now being used.

Accidents caused by industrial robots

To identify the dangers of industrial robots, the experience gained by reviewing past accidents is very helpful. Unfortunately, industrial robots have been in use a relatively short time. Thus, there is very little data available that clearly identifies accidents involving robots. Workplace accident reports involving robot use are likely to be buried deep in the classification of accidents caused by general power machines and other machines. But from available accident data, the following examples are instructive:

- A worker who was about to explain the operation of welding robots to his two colleagues was pinned by the machine as he stepped into the positioner area. A colleague operating the robot had pushed the start button without first making sure his co-worker was out of the operating range of the robot, thus pinning the victim's body between the positioner and the robot arm.

- A worker who discovered a defective piece on the conveyor line climbed onto the belt whilst the robot autoloader, a variable sequence machine, was stationary. The worker grasped the piece with both hands as he stood in the path between the line and centering machine that fed work onto the line. When the limiting switch engaged, the robot arm moved out to the conveyor, crushing the worker to death.

- A worker who noticed a problem with a planing machine as it operated in conjunction with a robot, first turned off the robot interaction switch and performed the normal confirmation routine with the manual operation switch. After adjusting the planing machine and turning on the robot inter-action switch the waiting robot arm flew out from behind and crushed the worker.

Each of the examples is a tragic robot-related heavy accident. But they serve to demonstrate clearly that robots have arms powerful enough to kill human workers, whilst suggesting some accident-potential operation patterns peculiar to robots. In each case, the accident occurred because the robot commenced operation without the victim's knowledge, who in turn had entered the dangerous area with little concern since the robot was idle. These accidents suggest that even at slow operating speed, robot arms can crush a worker as they move from outside the worker's line of sight, catching the unaware victim from behind.

Danger space

With some conventional machines the concept of a dangerous work zone does exist, but this zone is usually located inside the machine. By contrast, one characteristic of the robot is that a large surrounding area can be potentially dangerous. Even when the danger space overlaps the working zone of human operators, as with cranes for example, conventional machines are always under the control of humans and the accident causes are nearly always traceable to operator errors or to erroneous handling of the work to be done with the machines involved. However, with robots the danger space is anywhere within the arm's reach. In addition, the danger space would extend over a wider area should the robot lose control of the workpeice it holds.

The term 'danger space' is simply defined by the Japanese Industrial Standards as "the zone where a dangerous condition may arise when entered." A clearer determination of this danger zone can be see in the West German proposal to the International Organisation for Standardisation (ISO), which is

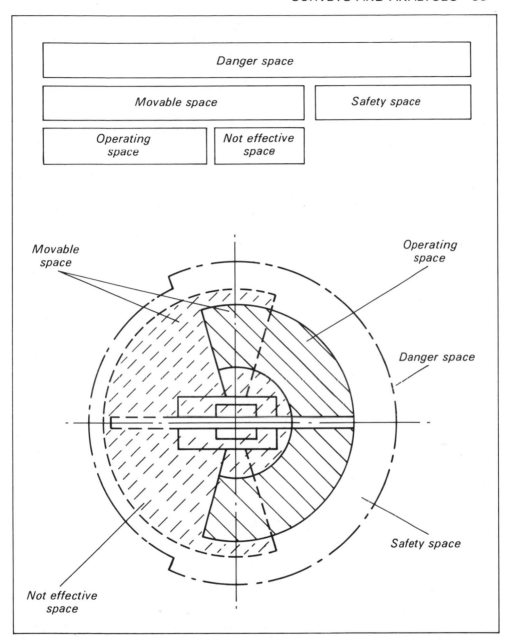

Fig. 1   Definition of danger space for industrial robots

working toward international standisation of terminology of industrial robots. As shown in Fig. 1, the danger space is defined as the area anticipating a safe margin within the periphery of the operating space, which includes the space covered by the moving parts of the robot. If a protective fence or some other protection is set up to screen the danger posed by the conduct of work within the danger space, or if there is apprehension about conducting work there, the area is defined as a danger space. In this sense, the West German proposal is very persuasive.

The following four incidents were provided from accident surveys[1] conducted by JIRA:

- A worker's fingers were caught between the workpiece held by a robot and the cutting jig, trapped by the up-and-down action of the robot's normal operation.
- A worker was cut by a thin steel plate held by the robot as it passed the plate over his hand during normal operation.
- The arm of a robot hit a worker's body during the normal operating sequence, bruising the worker.
- The robot arm moved out of sequence during manual operation, cutting the operator on the head as he was trying to regulate the robot arm.

During investigation for this report, several other examples of robot-related accidents were found, indicating that the incidence of such accidents is probably much higher in reality.

A similar survey of accidents was conducted in Sweden.[2] There, a survey of robot-related accidents over a 30 month period from January 1976 to June 1978 was made in the steel industry. The survey found the industry, which at that time used about 270 robots, had reported 15 accidents in the 30 month period. While the accidents were not described in detail, the survey analysis concluded that they were typical of accidents caused by robots, involving workers who carelessly got too close to the machines while they were operating, or workers who operated robots incorrectly by pushing the wrong control buttons, or being cut by tools grasped by robots. (For robot accidents in Sweden from 1979 to 1983, see page 49).

Latent dangers of industrial robots

To analyse the conditions which have the potential of causing or contributing to workplace accidents, and to evaluate the degree of danger, the most direct method is to statistically process or analyse the data of actual accidents. But because sufficient data to permit statistical processing are not available on industrial robots, this report attempts to evaluate the dangers of man-robot working environments, with the findings of workplace surveys at plants where robots are used as references.

Man-robot working environments

Human workers who deal with robots, must move, install, prepare, adjust, and maintain them as they perform their automatic work sequences. Of course, humans also program or 'teach' the work sequences to the robots. An attempt is made here to classify the operations and to access the potential dangers of working with industrial robots.

Moving and installation. The centre of gravity of a robot is not always at the centre of its body, and changes according to the position of its arm. Some robots cannot even stand upright until they are anchored to the floor with bolts. (It can be safely said that a machine that could not stand upright without anchor bolts was never seen on a shop floor in the past.) Extensively articulated robots are very unstable when being transported and are very difficult to handle if the arm is not locked. These stability factors should be considered whenever industrial robots are to be moved.

Preparation and adjustment. The adjustment and positioning of the limit switch, shock absorber, brake and other elements to install the robot in its working position often requires personnel to work in high places, balancing as they calibrate the device and running trial-and-error test operations. This kind of above-the-floor work is rare with conventional factory machines, and can therefore be considered peculiar to robots.

Operations to change the hand (grasping element) or connect the robot to form a system with other machines, or to install a guard fence are not seen to be much of a problem when they are performed with the power supply to the robot turned off.

Operations such as gain adjustment, offset regulation or auxiliary circuit regulation are conducted to make the fullest use of robots; what should be recognised, however, is that these operations have inherent dangers.

When the work sequence performed by the robot is to be revised, the pre-recorded cassette tape must be re-recorded after the new work sequence has been programmed. This work includes confirmation of whether the new program called for by the replacement has actually been set. But in some cases this confirmation work may be entrusted to a part-time worker. In such a case, there is always the possibility that untrained programmers might install an entirely unrelated program in the robot.

Programming (teaching) work. Programming is the work of preparing in advance the sequence of operations to be performed by the robot. Various sytems are available:

- Programming by numeric control language.
- Pinboard method.
- Combination of potentiometer and sequencer.
- Teaching box.
- Direct teaching by actually moving the arm of the robot.

In all of these methods, the work has a very important relationship to workplace safety, because it involves control of movements of the very powerful and potentially dangerous arm. Among the programming methods, the teaching box and direct manipulation are operations peculiar to the robot. The teaching box method of robot programming is a job involving a close human-robot relationship, in the sense that the robot is taught its work by numerous push-buttons operated by a human. This teaching work poses particular problems concerning safety. For one thing, the step-by-step nature of the method is likely to be accompanied by mental fatigue.

If the robot program cannot be easily changed with a simple action, such as pulling a pin as in the pinboard method, the programming is less dangerous since it helps safeguard against abnormal operation of the robot. It is necessary to place the robot under appropriate control so the program or the set position on the control panel cannot be changed indiscriminately. Similar safeguards are required for the potentiometer and sequencer method. In the so-called playback robots, programmed by teaching box or manipulation, appropriate control must be provided to make sure that the speed control will not be moved unexpectedly.

Test sequences. Supervision and testing the robot's performance, involving the 'check' and other functions, is conducted at low speed, or with no work load, or with no electric current flowing in the welding wire in arc welding robots. Thus it is impossible to fully test the programmed robot operations by simply switching them to automatic operation mode. Test sequences involve correction of errors and regulation of electric current. A close man-machine relationship is formed in this test operation work in the sense that the eyes of the operator must be keen to determine with accuracy when the robot is moving at a steady speed. Thus, the danger in performing supervisory test sequences must also be studied to ensure safety in this operation.

Starting operation and related functions. Before the robot can be used, various set-up operations must be performed: turning on the power supply and the oil pressure pump and manoeuvering the robot arm back to its starting position. Then, the automatic sequences to check the oil pressure, oil temperature, abnormal start and other steps related to the initial operation of the robot can be checked to determine that the robot is in the condition of 'preparations completed.'

The robot is then ready to be switched over to 'automatic' operation. When performing these steps, it is important to confirm with certainty that there is no dangerous condition within the operating zone of the robot. As past accident cases have clearly demonstrated, a careless or haphazard approach to the set-up procedures is an invitation to disaster.

Steps related to automatic operation. While the robot is operating, the human partner must mount and dismount the work, dispose of cutting waste of lathes and other peripheral equipment or change cutting tools, transfer the work to a conveyor, inspect the work or monitor productivity.

These operations are not directly related to the functioning of the robot, but the zone of the robot arm if the robot is a component of a production line ceiling-suspended are typical, requiring the appropriate safeguards to protect the human workers from robot dangers.

In the chip dressing process, although performed when the robot is stopped, the overall operation is performed repeatedly, and can thus be considered a typical process in the automatic operation of a welding robot.

Unexpected work halts. Robot stoppages can be categorised as follows:

- Emergency halt with the emergency stop button.

- Temporary halt with the pause button.
- Malfunction halt due to detection of an abnormality.
- Runaway halts for machine failure.
- Condition wait halts for machine recycling.
- Work termination halts at the end of a work program.
- Apparent halts due to fixed-point position control.

Among the ways to stop robots are: cutting off the power, cutting off the drive and oil pressure pump power supply, cutting off the power supply for servo control, or pauses. As Table 1 shows, the robot design philosophies differ widely on which halt procedure is best used for the particular halt conditions. The emergency halt, in recent years, has been done with a dynamic brake.

In this area, the conventional concept of turning off the circuit in an emergency, as in conventional machines, no longer applies. Thus, robots have many unique features in which conventional safety technologies do not apply. It is necessary, therefore, to develop new safety procedures for robotics.

When a robot is sophisticated, many sensors monitor the control and drive systems to ensure that it will stop automatically if a malfunction has occurred or if any abnormality is detected by its peripheral equipment. Emergency halts are the intentional actions of humans to stop the robot when a malfunction or other condition has arisen but was not caught by this detection net, including the failure of the detectors themselves. There have been cases where the operator caught in the positioner of the robot has been rescued by emergency halts. Thus, the emergency stop function is very important as an ultimate safety measure.

In most manual halts, and some automatic halts, the operator may come within reach of the robot arm while investigating a problem or making adjustments. Accidents caused by a sudden, unexpected starting are quite possible under such circumstances.

When air pressure-activated control arms are caught in peripheral equipment, trouble-shooting or repair should never be undertaken until after the air pressure has been released. This simple safety measure applies equally for robots and for conventional machines.

Just because a robot has stopped, it cannot be assumed that it is safe to work

Table 1. Types of robot halt and their conditions

Type	Power supply break-off	Cut-off of drive power supply	Cut-off of servo power supply	Pause	No particular condition
Emergency halt	●	●	●	●	
Temporary halt			●	●	
Abnormal halt	●	●	●	●	
Run-away halt					●
'Condition wait' halt		●		●	
Work-end halt	●	●	●		
Apparent halt				●	

on it. The halt may be a temporary, machine-induced pause, as when the workcycle ends. As the example of accidents given above illustrates, an easy-going attitude toward approaching a halted robot is very dangerous. Since large production lines are the most difficult to stop, this might explain why a worker would perform such an unsafe act.

Repair and maintenance of robots. The repair and maintenance of a robot or peripheral machines, in many cases, must be performed within the reach of the robot arm, or in a very tight space in a very unnatural, awkward posture. As the robot is a machine with a heavy metal arm, safety must be stressed even more in the repair and maintenance of robots than it is for conventional machines.

Reliability of industrial robots

The most important consideration concerning reliability is the danger of the robot making unexpected movements through a failure of its control system or through electrical interference – the so-called 'runaway danger'. Another problem is the danger of the robot releasing the work it is holding.

An automobile assembly plant in Japan[3] has compiled the problems it has encountered with robots. At the facility, the following accidents occurred:

- In a training course using a robot, the slewing shaft suddenly swung away from its preprogrammed direction.
- The arm of a robot suddenly shot up as the oil-pressure source was cut off after the robot ended work.
- A robot made a motion that was not part of its program.
- A robot started moving as soon as its power source was switched on, although its interlock conditions were still not read.
- When operating alone, a robot destroyed the work it was to weld because of a mistake in program instruction.
- When the power switch was turned on, the robot arm sprang out and the slewing shaft rotated and stopped after becoming entangled in the welding machine.
- During hot summer weather, the arm of a robot suddenly sprang up, although it had otherwise been working normally.

Malfunctions, such as the unexpected start, reportedly occur at a rate of several dozen annually at the factory. The most common causes are reported to be electrical noise, problems with pressure valves and servo valves, encoder-

Table 2 Frequency of robot problems

Problem	Frequency (%)
Faults of control system	66.9
Faults of robot body	23.5
Faults of welding gun and tooling parts	18.5
Runaway	11.1
Programming and other operational errors	19.9
Precision deficiency, deterioration	16.1
Incompatibility of jigs and other tools	45.5
Other	2.5

related trouble, pcb malfunctions or abnormalities and errors traceable to misjudgement or erroneous operation by human workers.

Table 2 charts the frequency of robot problems[4]. Failures in electrical circuits, including the microcomputer, account for the largest number of cases, and malfunctions of the jig and other instruments come next. Runaway is defined as an unreproducible erroneous action caused by short-circuits, noise or similar interference. At 11.1%, it is by no means a small portion of the total statistics.

Table 3 Mean-time between failures (MTBF)

MTBF (hours)	Frequency (%)
Under 100	28.7
100– 250	12.2
250– 500	19.5
500–1000	14.7
1000–1500	10.4
1500–2000	4.9
2000–2500	1.2
Over 2500	8.5

As Table 3 indicates, the mean time between failures (MTBF) is rather small: 75.1% of robots have an MTBF of 1000 hours or less and those with a MTBF of 100 hours or less account for 28.7%.

Some robot manufacturers claim that robot reliability has been greatly improved. But even with improvement, the best MTBF yet achieved is 2500 hours. When such machines are operated by people and are formed into systems with other machines, they inevitably inherit a considerably higher incidence of problems. Under such circumstances, it should be fully recognised that robots have not yet attained adequate reliability.

Identification of accidents caused by industrial robots

Accidents occurring in industry in general can be broadly categorised into the energy-conversion type and the energy-loss type. Accidents involving robots can be classified as accidents of the energy-conversion type. In the human-robot working environment, accidents arise because the energy of the system is mischanelled into a form that injures people.

Table 4 illustrates the types of accident that can be expected in work involving the robot, as indicated by open circles; work which is thought to have a particular marked accident relationship is indicated by solid circles.

Since robots can be used in many ways, it is conceivable that they may be used to handle explosives, dangerous and harmful materials, high-temperature or low-temperature objects, radioactive materials, and so on under a controlled environment. The dangers inherent in the use of robots obviously change drastically according to the form and mode of operation.

Table 4 Accident identification

Energy source	Type of accident	Robot related work with potential danger							
		Transport and installation	Grading	Programming	Test running	Starting	Work attendant to automated operation	Maintenance to eliminate malfunction	Maintenance and repair
Potential									
Position	Collision with robot	○	○	○	○	○	○	○	○
	Fall	○	○						○
	Hit by falling object	○		○	●	●	●	●	●
	Hit by toppling robot				○	○	○	○	
Electric	Electric shock		○	○	○	○	○	○	○
High-pressure	Rupture			○	○	○	○	○	○
Mechanical	Hit by thrown object				●	●	●	●	●
	Collision and hit			○	●	●	●	●	●
	Caught between robot and other object	○	○	○	●	●	●	●	●
	Cuts, scrapes, tears	○	○	○	●	●	●	●	●
	Caught in robot		○	○	○	●	●	●	○
	Cuts		○	○	○	○	○	○	○
Chemical and biological	Explosion								
	Contact with dangerous and harmful substances								
	Exposure to sonic wave strain								
Thermal	Contact with very hot or very cold object	○	○	○	○	○	○	○	○
Radioactive	Exposure to ultra-violet or infrared rays								
	Exposure to radioactive rays								

Potential energy

Position energy. In the human-robot working environment, there is always the possibility that the human worker may fall from a high place, losing balance or tumbling. Although not listed in Table 4, this potential danger constitutes an important safety problem, especially with large, ceiling mounted (X-Y coordinate) robots, and requires adequate study before their use. The stop position of the area is sometimes located high up and, in such cases, the danger of accidental falls is always present.

With general factory machines, the possibility of an operator falling because of slipping or tumbling is an ever-present danger. But with robots, the possibility of the machine itself toppling is an added element of danger, especially since its singleside support construction means its centre of gravity is not fixed. Thus it is always possible that the robot may fall when it is being installed, or when it starts or stops, or when it is overloaded. And in Japan in particular, the possibility of natural disasters, such as earthquakes, is an added element of danger that must be considered.

Also, the danger of the work held by the robot slipping and falling should be taken into consideration.

Electrical energy. Electric-shock accidents caused by short-circuits in the wiring or connectors of the robot are common to other machines. Since robots often have many movable arms, there are usually other cables to distribute electricity to, for example, spot welding tongs, welding holder, welding wire, or any other tools at the tip of the robot hand that, because of their constant movement, are subject to wear and tear and rapid deterioration, leading to frayed cables and electric shock.

Even if the operator does not receive a shock, there is the possibility of power failure or short-circuit that could cause erroneous action of the robot. Also, electric sparks can cause fires or explosions.

High-pressure energy. There is a chance that an operator may be hurt by 'whipping' hoses that could break connection or rupture due to oil-pressure or air-pressure fluctuations.

Mechanical energy

If a robot is overloaded, or is being used to handle work sizes and shapes for which it was not designed, the object may fly off and hit an unsuspecting worker. Research[5] has shown that it is quite possible for work to fall from the robot's grasp if it clings to the surface of the grasping device. The hazard of the robot releasing its work can also occur easily if the robot arm collides with something or when it comes to a sudden stop. The potential for collisions, scratches, cuts and abrasions from the robot arm-joints is always present any time the operator is within the danger zone.

The prospect of a two-pronged attack, trapping the worker between the robot and its peripheral equipment, is multiplied by the chances of construction flaws in the robot, especially in robots that have a link mechanism. This link mechanism contributes to the risk of accidents because it

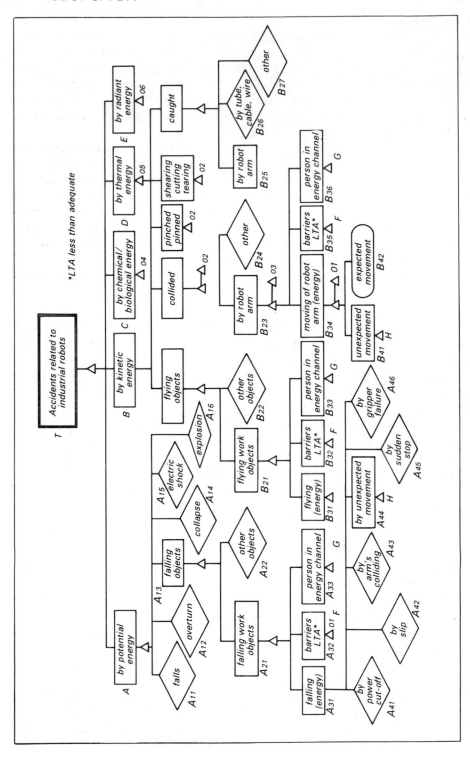

Fig. 2   Fault-tree for robot-related accidents

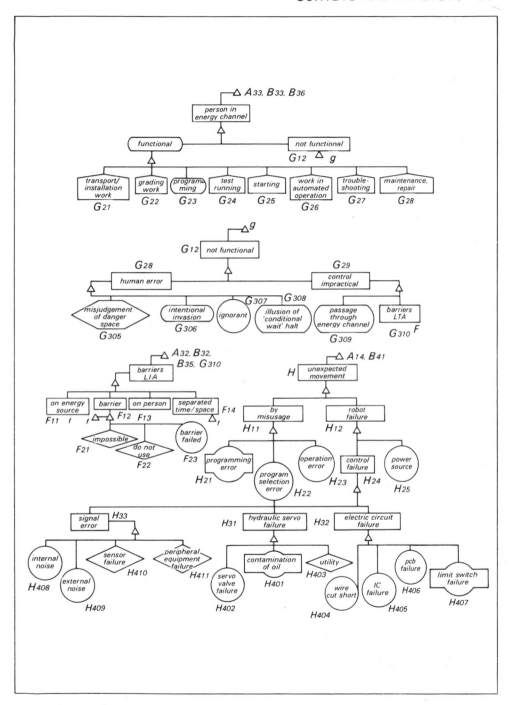

Fig. 2 continued

cannot be covered. It should always be kept in mind that the robot is part of a large system, which multiplies the danger of a worker being pushed or dragged into the path of a lathe or other machine on the line.

Chemical and biological energy

Robots can present the risk of fire or explosion when they are used in an explosive atmosphere or near a combustible substance, such as when painting. The fire hazards might be electric sparks or mechanical friction. Robots that are used to handling dangerous or hazardous objects or materials present added hazards resulting from malfunctions or operator errors. The human element must also be considered, e.g. backache, numbness, hearing difficulty and other disabilities caused by improper maintenance of the robots or lack of consideration of the operations to be performed by humans.

Thermal energy

Robots that perform casting, forging, heat treatment and other such work, can injure workers who come into contact with hot materials. A similar risk exists where the robots handle very cold materials.

Radioactive materials

Robots that handle radioactive materials or emit infrared or ultraviolet rays can expose operators to the harmful effects of these energies.

Study of robot accidents by fault-tree analysis

All of the potential dangers in the man-robot work environment are the result of combinations of unsafe conditions and unsafe actions. A deductive analysis of the conditions for combination of these factors and their cause-and-effect logical constructions can be made by fault-tree analysis (FTA). This method[6] is a widely used tool for deductive analysis; Fig. 2 shows an analysis of the occurrence of accidents in the man-robot workplace.

In this study, the FTA concentrates on accidents arising from characteristic functions or structural features of robots. Since robots have a wide range of applications, the work handled and tools used are also diverse, as are the forms of energy handled. For this reason, a full study of accidents involving thermal energy, chemical and biological energy, and radioactive energy, which are not discussed sufficiently here should be undertaken to include countermeasures to ensure safety from these dangers.

From the FTA study of the dangers inherent in the man-robot work environment the following ideas are offered as countermeasures for making industrial robots safer:

● *Reduce the chance of abnormal energy conversion.* This can be attained by improving the hardware reliability of robots and by enhancing the reliability of the people who operate them. The latter may require a system of exclusive operators properly trained for the job, as well as improving and standardising the engineering designs of robot control panels.

It is not always possible to disconnect the robot's power supply when an emergency halt occurs. As long as the power supply is alive, the danger of the robot 'running away' exists. It is therefore necessary to study ways to ensure complete reliability of the emergency brakes of robots. In addition, robot operators should also think of ways to reduce the amount of work required within the danger zone.

- *Minimise abnormal energy conversion.* This means reducing the number of operations requiring operators to approach their robots by lowering the operating speeds of robots. With constant surveillance of the loads and halting the robots whenever loads vary in weight, size or shape from the specified work load and other load-orientated motion studies, the chance of erratic operation or malfunction can be reduced.

- *Optimise defensive measures.* As mentioned earlier, all kinds of accident-prevention measures should be taken. Among the physical devices and equipment that should be installed are safety fences, safety plugs and mechanical brakes. Sensors to detect presence of human workers (photo-electric, electrostatic, supersonic, laser and visual monitor sensors), as well as machine malfunction detectors, should also be employed.

- *More precise robot control.* Whilst being aware of the dangers of robots, it is important to thoroughly educate operators, programmers, maintenance and repair engineers, as well as part-time workers, on appropriate safety precautions.

Since the actual extent of accidents caused by robots is not known, and since there are no formal standards or rules regarding safe operation of robots in the workplace, the development of such standards is being left largely to the subjective policies of industrial robot manufacturers and users.

In summary, safety measures when using robots are of pressing necessity for the following reasons:

- Since robots are automated machines, they are often regarded as having nothing to do with human workers.
- When programming, humans must enter the robot operating zones and cannot do their jobs if the power to the robots is cut off.
- Monitoring, tool changing, chip grinding, inspection and other operations involving robots and their peripheral equipment must still be done by humans.
- Adequate reliability of robots has not been assured.
- Since each robot installation is different, each presents unique application problems; it is impossible for robot makers alone to ensure that robots are essentially safe.
- To correct problems with peripheral equipment, it is necessary to enter the robot danger zones.
- In many cases, experience must substitute for data concerning accidents within a factory.

As the use of robots has increased and their application has grown, advances have been made in robot design and manufacture. The development of technology for intelligent robots in particular can be expected to advance rapidly. As that happens, robots will perform actions that may be completely unpredictable by humans. As the uses of robots become better known, their degree of sophistication can be expected to increase, making the issue of safety in the human-robot work environment a matter of extreme importance. Only when robots themselves are able to detect the approach of humans and perform appropriate actions to avoid accidents, will safety in the human-robot workplace be assured.

References

1. Survey and Study on Standardisation of Industrial Robots, No. 2, 1976. Japan Industrial Robot Association, Tokyo.
2. Carlsson, J. 1974. *Industrial Robots and Accidents at Work*, TRITA-AOG-0004. Arbetarskyddsstyrelsen, Solna, Sweden.
3. Shima, S. 1982. Safety control on introduction of industrial robots to factories. *Safety*, 33(3): 18.
4. Sato, K. 1982. Case study of maintenance of spot-welding robots. *Plant Maintenance*, 14(3): 28.
5. Sugimoto, N. 1981. Safety design of vacuum pad. *Mechanical Hands*, 12.
6. Johnson, W. G. 1973. The management oversight and risk tree, MORT, SAN821-2. US Atomic Energy Commission.

3
System Design, Implementation and Methodology

This section examines the design of safe robot systems in very broad, top down terms. It looks at safety systems for the future, at real-time risk assessment by a separate safety computer, at designs for safeguarding, at CAD as a robot workplace design tool, at a safety watchdog computer which protects the robots, at the hardware and software of the robots themselves, and how these are designed with safety in mind. Finally, it examines training requirements for the safe and effective implementation and operation of industrial robots.

The first paper looks at the safety requirements of an automated production system. It is assumed that the system, which may range from a single CNC machine to complete flexible production cells or automatic production systems, will be computer controlled. But an automatic production system probably including robots is decidedly different from a traditional manufacturing system and the adoption of a classic safety approach may interfere, often to no purpose, with the functioning of the plant. Bonfioli et al. propose a system of on-line real-time risk assessment performed by a separate safety computer. Although this would not be acceptable to most current guidelines, the concept is important because it raises issues which will need to be faced sometime. The proposal, together with the proposed experimental work, point a way to the future.

Bellino, in his paper 'Design for Safeguarding', discusses a design approved to help prevent accidents occurring during the programming/teaching mode or during the trouble shooting/maintenance mode. These accidents may be caused either by equipment failure or human error. The paper presents a systematic but general design procedure which together with a functional checklist will prove useful to design engineers as a framework within which to consider their organisations' specific problems.

In the next paper the Editors and Nick Taylor discuss the use of the GRASP computer aided design system to plan and evaluate robot workplaces. Some specific safety routines are discussed, but more important is the message that CAD is a flexible and natural way to plan workstations. In particular it enables designers to locate the items of the workplace, plan the work tasks and check that collisions do not occur. The planned work task then becomes, suitably post-processed, the robot program produced off-line.

The fourth paper by Roger Kilmer and colleagues from the US National

Bureau of Standards discusses the use of a standalone microcomputer system to monitor robot-related operations at the workstation so as to detect unusual conditions and to stop the robot before it can damage itself. This is robot safety with a slightly different meaning! Included in the system are forbidden volume routines which can be used to impose restrictions on the tool plant. Links can be seen between this paper and the safety system proposed by Bonfioli et al. and the use of CAD to design robot workplaces discussed in the previous paper.

The inter-relationship between hardware and software is clearly illustrated by Barbara Cook of IBM. In particular software is used to enhance the standard machine safety hardware and so increase the protection against personal injury and equipment damage due to hardware failure.

This section closes with a paper by Lloyd Carrico which puts it all together. The point is made simply that the first and most important consideration should be the familiarisation and training of all personnel. Once trained, the users' staff can, with the vendors assistance, design an effective and safe robotic application. The next stage is the implementation where the author suggests that most safety factors are overlooked, particularly if changes to the design are required. In addition to the discussion on training there are some good practical safety hints, e.g. the rear loading of machine tools to reduce danger to maintenance personnel, and there are relevant comments on perimeter guarding, light curtains and pressure sensitive mats. This provides a useful link to the papers on these particular topics discussed in Section 4.

Safety System Proposal for Automated Production

M. Bonfioli, C.F. Marcolli and C. Noé
Politecnico di Milano, Italy

With the development of CAM and the introduction of FMS, the problems connected with the interactions between man and machine or machine and machine take on great importance. It is intended that this paper will contribute to the development of safety systems which are able to exploit the possibilities given by CAM. The authors outline their proposal for a safety system which is able to analyse all the situations in the work process and, according to the results of the analysis, to take appropriate actions in order to maintain the highest level of safety with the lowest loss of production.

The changeover from production systems made up of single machines, at most gathered together in departments, to complex production systems made up of groups of machines with various logical, physical and functional connections, has brought about a thorough re-evaluation of the problems related to the design, management and operation of a production plant.

The approach to safety problems for plants using conventional machines has remained substantially unchanged over recent years. Where man–machine interaction is indispensable, the prevention of accidents both in respect of the machine and the operator is, in practice, entrusted to the attention of the operator. Where man–machine interaction is not indispensable, the prevention of accidents to personnel is obtained by the physical separation of machine and operator, while the prevention of accidents to the machine is based on the assumption that as the machine always and only carries out the same operations, under the control of highly reliable mechanisms, such operations will not generate situations in which the machine is endangered. This approach has allowed the construction, in satisfactorily safe conditions, of productive structures, whether they be single-purpose production lines, groups of highly automated machines, or manual machines.

However, with the advent and distribution of production systems managed by computers, both single CNC machines and production systems or flexible production cells, safety problems have taken on a decidedly different dimension. Indeed, if we consider a CNC machine tool, the possibilities of programming errors or electronic faults imply a whole new series of accident risks which, because of the way they manifest themselves, are not typical of the cases occurring with manual machines or traditional machines.

As a result, problems related to interactions, and consequently collisions of moving parts of the machine and other parts (moving and/or fixed) belonging to the machine itself and external to it, have taken on great importance.

In integrated production systems, above all, such interactions even occur between different, separately controlled machines, and the problems become even more complex.

Furthermore, the classic definition of machines as being 'open' or 'closed' with regard to the operator is now substantially useless, given the different modality of use of such machines.

From the point of view of the operator's safety, the ideal procedure, given the dangerous nature of accidents caused by collisions, would be to 'close' the machine or complex of machines, accessories and slave units. This, however, would make it impossible to limit the damage in case of programming errors or electronic faults, above all in the setting-up phase of the work programme, because of the inevitable delay to operator intervention. In addition, such a measure would make it more difficult to maintain one or more of the components of flexible production systems without stopping the functioning of the whole system; it appears to us that it can therefore be proposed only in the case of routine operations, that is, when there is little chance of collision because of the certainty of the correctness of the work programmes. On the other hand, the safety measures typical of manual machines, such as limit switches, appear totally inadequate for numerically controlled machines because of the limits they would place on machines which are, by their very nature, extremely versatile.

Safety system proposal

The safety requirements of an automatic production system are decidedly different from those of a traditional system. On the other hand, given the delicacy of management of the most highly developed automatic production systems (e.g. FMS), it is evident that the classic types of anti-accident apparatus, based substantially on consensus signals of the on-off type, can interfere, often to no purpose, with the correct functioning of the plant.

The safety system proposed is capable, unlike traditional systems, of managing the different degrees of danger that may appear through a series of suitable interventions. To obtain this result, the whole system is based on the presence of a mini- or microcomputer, which enables complex analyses of critical situations to be carried out in real-time. One could think of using the production computers to manage safety as well. However, the problems relating to the reliability, and rapidity of response of the safety system, and above all, the difficulty of realising software which would not provoke reciprocal interference between the two sectors (production and safety), are such as to make physical separation of the safety computer from the system of production management advisable.

The principal characteristics which qualify the system are discussed below:

- *Intrinsic safety.* The system must, in the case of a fault, be capable of signalling it, removing it if possible, and failing this, behaving as if the faulty apparatus gave rise to the condition of maximum localised danger.

- *Selective protection.* The system must be capable of recognising all dangerous situations, of classifying them according to the level of danger, and of carrying out immediately all and only the actions necessary to remove the conditions of danger, while interfering as little as possible with the production system. To do this the system must be autonomously capable of: recognising whoever or whatever enters a forbidden zone, foreseeing the results the intrusion could give rise to in the worst possible case, checking if the intrusion is prescribed by the production system, taking the most appropriate measures for removing the dangerous situation, checking that the dangerous situation has in fact been removed, and informing the management of the production system of the situation which has arisen.

- *Possibility of partial disconnection.* The system must be informed of the possible necessity of operating in conditions of limited safety or conditions different from those prescribed (e.g. during setting up or maintenance) and must be capable of providing the required new level of safety.

- *Recording dangerous situations.* It is essential that dangerous situations should be recorded and analysed in order to achieve the objective of preventing their recurrence.

- *Integration with the production computers.* The safety system can carry out a certain number of interventions without acting directly with the computers of the production system. Such interventions could be activation of alarm signals, putting machines out of action, closing barriers or protective devices, etc. However, other interventions, such as the reduction of speed or the suppression of certain dangerous operations require a 'dialogue' between the safety computer and the production computers.

Functional characteristics

The proposed system has an interface with the environment which allows all the information related to the functioning of the production system to be provided to the computer which manages this information.

In the most general case, the production system will be made up of a group of operating machines (CNC machine tools, robots, automatic transport systems, etc.) each of which is managed by a mini- or microcomputer which will be called a governor unit (GU). All the GUs are interfaced with a plant computer that permits the management of the whole system. The logical links between production system and safety system can vary according to the degree of complexity of the interventions by the safety system itself. The most simple connection is shown in Fig. 1.

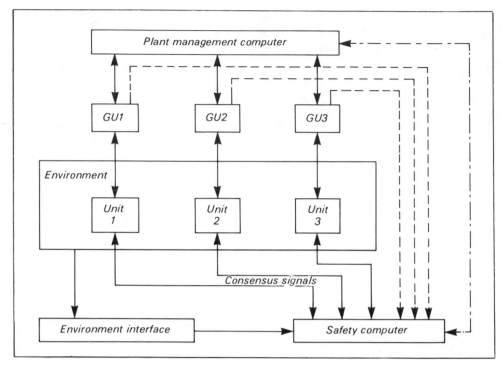

Fig. 1   Logical connections between the plant and the safety system: the connection between GU and safety computer can be either direct (broken lines) or indirect (stippled lines) via plant management computer

As can be seen, the interface with the environment informs the safety computer about the situation inside the production system. Should dangerous situations arise – either human invasion or risks of impact between machines – the safety computer sees to the elimination of the consensus signal on the operating machines involved and thus stops them.

The logical links in the system extend to include the GUs. In order to effect this, it is possible to realise a direct interface between a GU and the safety computer or to pass through an indirect connection by way of the plant computer. From a functional point of view both solutions are equivalent. The second scheme allows intervention, should dangerous conditions arise, by limiting the kinematic parameters of the operating machines, but without going as far as stopping them. Only in cases where the dangerous situation can not be removed in this way should the safety computer intervene by inhibiting the consensus signals and thereby stopping the machines directly. All this obviously leads to less interference by the safety system with the production system.

Solutions – advantages and disadvantages
Unlike traditional safety systems which divide the environment in which the machines operate into two zones (one of free access to any external factor, the

other totally shut off), the proposed systems determine, moment by moment, the area of potential danger (defined as the site of the points in which a contact could occur in case of accidents) and checks that there are no intrusions in this area.

With the changing of the working conditions – position and speed of the various moving parts –the areas accessible to external elements also change, and thus, during operation and in conditions of safety, areas of operation not being used by individual machines are rendered usable.

Interactions between different machines and between man and machine are therefore possible with the maximum operative and safety levels. Furthermore, since the area of potential danger is directly proportional to the response time of the system, it is possible within certain limits, to reduce this area by operating on the software of the safety computer.

Assuming that the conditions of danger in the various areas controlled are scanned cyclically, it is possible, by modifying the scansion, to operate more checks per unit of time on the areas in critical conditions (obviously at the expense of checks on other areas).

By way of an example, suppose we have to control m areas of the system and that the checking time is constant and equal for all of them. Starting from a system in which N checks are made of each area in the unit of time and, for example, doubling the checks on one single area, there remain $N' = (N(m-2)/(m-1))$ checks per unit of time for each of the remaining areas.

Operating in this way facilitates, even with the first configuration proposed, the limiting of interference by the safety system, as this is able to pay greater 'attention' just where it is required.

Compared to traditional systems, the systems proposed involve higher investment and greater complexity of realisation, to be set against the notable operational flexibility offered. The safety computer, in particular, needs particularly sophisticated software, and the environment interface does not appear to be easily and quickly realisable, given our current level of knowledge.

Fig. 2 shows the block diagram related to the operational logic of the safety system proposed.

Operation of the safety system

The system, as has already been seen, is based on the knowledge, moment by moment, of the movement characteristics of each object situated in the area of work of the machine(s) and on the determination of the space required for stopping in the worst hypothetical case of loss of control. Let us start by considering the most simple case of a part of a machine – which can be indicated schematically by a point – endowed with a single degree of freedom which consists of the possibility of movement in a straight line.

With reference to Fig. 3, assuming that v is the instantaneous speed at the moment of loss of control, a is the maximum acceleration which can occur with this machine, R is the maximum time of intervention of the safety system, and d is the minimum deceleration guaranteed by the intervention of the safety system, the space needed to stop, S, in the worst possible case is given by:

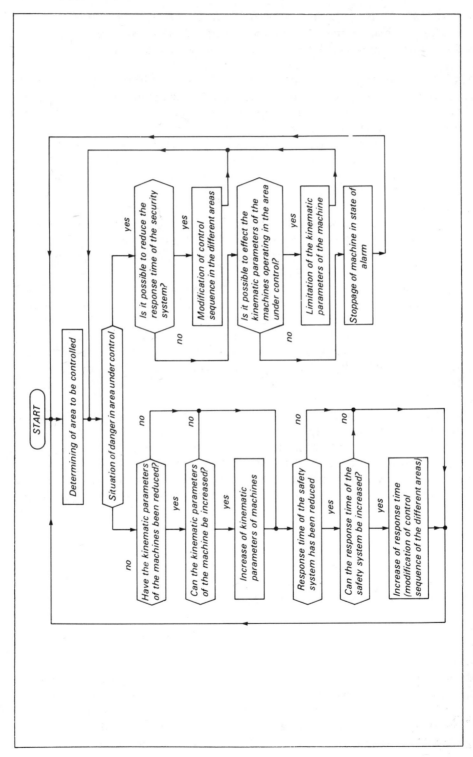

Fig.2   Functioning of safety system

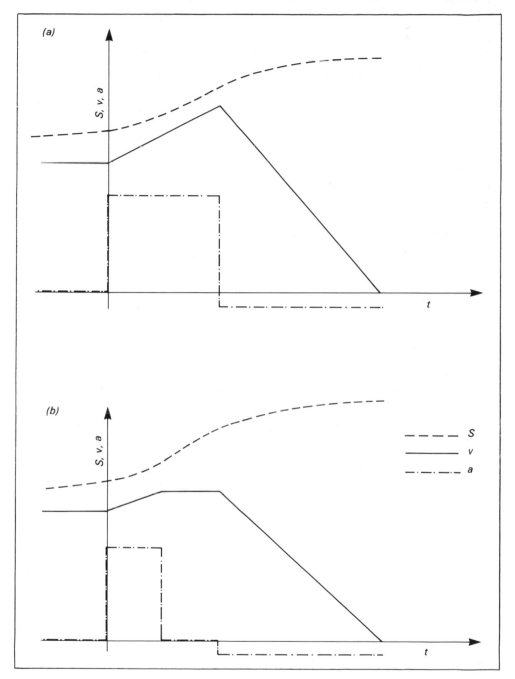

Fig. 3   Changes in speed and acceleration and distance travelled from the moment of the fault up to the stoppage. In the course of this period, maximum speed is reached (case a) or not reached (case b)

$$S = vR + aR^2/2 + (v + aR)^2/(2d)$$

if the speed attained after the delay R is less than or equal to the maximum speed obtainable by the machine, or if the maximum speed ($v$max) is inferior to the speed which can be obtained in the time R with acceleration a, S is given by

$$S = v(v\text{max}-v)/a + (v\text{max}-v)^2/2a + v\text{max}(R-(v\text{max}-v)/a) + v\text{max}^2/(2d)$$

which can be reduced to

$$S = v\text{max}.R + v\text{max}^2/(2d) - (v\text{max}-v)^2/(2a)$$

The space needed to stop, S, also represents the minimum safe distance. Now if, for whatever reason, this distance must be reduced, hypothesising a and d as being related to the machine and so not susceptible to modification by the safety system, it may be noted that S diminishes both as v diminishes and as R diminishes. In the event of having an 'intelligent' safety system at our disposal, it would seem convenient to reduce R as far as possible before intervening by limiting the speed of the machine.

A simple numerical example will show this convenience. With reference to Fig. 4, where $v = 0.5$m/s, $a = 10$m/s^2, $d = 1$m/s, and $R = 0.1$s, we have $S = 1.225$ m. Assuming we have an obstacle at 0.5 m it can be seen that the speed can be kept unvaried while operating in conditions of safety if the response time comes down to about 50 ms.

In the given example, at the approach of a dangerous situation, the safety system should be capable above all of increasing the control (and thus decreasing the response time) over the effected unit – and this is possible by reducing the control, in a way compatible with safety, over the other units – and only when the situation still appears dangerous, even with the maximum level of control, must the system intervene (if possible by limiting the speed of the machine, or if that is not possible, by stopping the unit). Obviously, if the safety system was capable of limiting the maximum acceleration as well, or rather, of increasing the minimum deceleration, it would be possible to manage more critical situations in conditions of safety without stopping the machine.

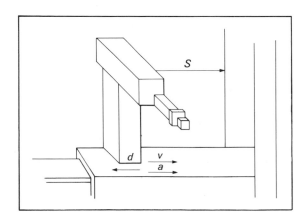

Fig. 4 In an element in rectilinear motion, the space required to stop, S, is a function of v, a, d and to a greater extent on the time of intervention R of the safety system

More complex cases can be examined with the same criterion which has been applied to this simple case. For example,

(i) Two or more elements in motion in the same straight line.
(ii) Two or more elements in rectilinear motion in the same plane on different trajectories.
(iii) Two or more elements in non-rectilinear motion in the same plane.
(iv) Two or more elements in motion on different planes.
(v) Two or more elements in motion in space.

Case (v), obviously, having general validity, includes all the other cases mentioned.

Given the necessity of having the shortest possible response times at our disposition, it appears evident that the algorithm of recognition of the situation of danger on the part of the safety system must be as simple and hence as fast as possible. This renders it necessary to work out appropriate algorithms for each individual case, in order to minimise the computing time.

Problems related to the feasibility of the system

The practical feasibility of a safety system of the kind proposed presents some problems which require consideration even in the logical conception phase. In particular, there are two most delicate problems: the realisation of the interface with the environment and the effective conformity of the safety computer to the prospective requirements. In order to reveal the presence of intrusions caused both by persons and by objects in the operating area of a machine, or rather in the area under control, it is possible, at present, to use both mechanical and electronic transducers.

The signal that is emitted by mechanical transducers is generally of the on-off kind, whereas that emitted by electronic transducers can be both of the logical or of the analogue kind.

In the field of mechanical transducers the most common, cheap and easily applied ones are limit switches and similar devices, and pressure-sensitive footplates.

The response of these transducers to an external stimulus is obviously of the logical kind, even if one could conceivably obtain signals variable in a linear way, depending on the pressure exerted, from the footplates. However, the margins of uncertainty of interpretation of the signal would be so considerable in this last case as to invalidate the meaning of the results and to make them unreliable.

The field of electronic or electric transducers is much greater; often the components have a high cost, but one which is negligible compared with the cost of a group of numerically controlled machines. These transducers make it possible to obtain more valid results in the field of protection. The most common and simple electric or electronic devices usable are photocells, photocell barriers, capacitive devices, microwave detectors and passive infrared ray detectors.

Photocell detectors give rise to a variation in their electric parameters which varies according to the light striking the sensitive element. Capacitive detectors, also known as proximity detectors, are generators of radio-frequencies, which vary the frequency generated with the approach of a body that introduces parasitic capacities into the oscillating circuit, modifying its work parameters. Capacitive detectors have some problems of instability of functioning which make their use in safety appliances inadvisable.

Passive infrared ray detectors pick up thermal disturbances of the field where they are caused by bodies which invade it. Their use is extremely difficult in an environment, such as the industrial environment, in which thermal disturbances are common.

Microwave detectors produce a signal of radiofrequencies of approximately 10GHz which can be directed and focused even at a considerable distance. When a foreign body which is not transparent to the microwaves is introduced into the wavepath, a signal proportional to the component of the relative speed (between the body and the detector) in the direction of the wavepath is generated through the Doppler effect.

In conclusion, at the present time, the transducers which may be said to be most technically reliable and immediately usable in a safety system, apart from those of the conventional mechanical type, are those of the optical and microwave type.

The choice of a safety computer is largely determined by the dimensions and complexity of the system to be controlled. It is, however, possible to recognise some common requirements which allow the choice to be limited to a fairly well-defined family of machines. The principal characteristics of a safety computer (by which we understand machines, hardware and software together) must be as follows:

- High speed, since as has been seen, the degree of safety of the system is inversely proportional to its response time.
- The need to auto-diagnose faults and malfunctions in order to realise the concept of intrinsic safety.
- The possibility of operating in a partially different mode from the standard one in order to allow access (e.g. for maintenance) to some machines only.
- Rapidity of repair or replacement of faulty elements.

Microprocessors with their vast range of accessories, seem to meet the stated requirements adequately. In particular, some cheap 16-bit microprocessors are available with execution of instruction times of the order of a few microseconds. Assuming that such machines are programmed directly in ASSEMBLER, with the purpose of optimising the programs, it is reasonable to assume that a single control loop can be carried out in a few milliseconds. In any case, processors with decidedly superior computing speeds are available, which, despite the fact that in addition to greater cost they pose some difficulties of use, may be considered for use in the most critical cases. Given the limited dimensions of the software, it could conceivably be directly stored in the central memory, using permanent memories.

The presence of an external memory appears, in fact, not to be necessary. At most it could be used to record and statistically analyse the interventions, if a connection to the plant computer was not planned.

The problems of reliability and auto-diagnosis of faults can be easily overcome by specifying two machines in parallel used for reciprocal control and enabled to function singly only in the case of the breakdown of one of the two.

The possibility of non-standard functioning can be obtained by acting through the plant computer or through a suitable console and must in any case be allowed for in the software development phase.

The problems of speed of repair can be resolved, as is now common practice, by subdividing the machine into immediately replaceable boards.

Software appears to be the most critical part of the system by far, and although its technical feasibility is not in doubt, it is difficult at present to evaluate its realisation economically and in terms of time.

Example of an integrated safety system

By way of illustration, the problems related to the realisation of an integrated safety system, conceived according to the philosophy described above, have been examined. The intention is to produce a safety system for the robot with two arms shown in Fig. 5. The two arms are controlled by the same GU.

Strictly speaking, the problem is a spatial one, however a satisfactory solution can be realised from the study of the problem in one plane. This involves an increase of the area of potential danger, but also a simplification of the calculation algorithms and hence a reduced response time.

From the kinematic characteristics of the robot given in Fig. 6, it is possible, if the characteristics of the movement are known, to derive, moment by moment, the area of potential danger for each arm (Fig. 7). In fact this area is limited by the furthest positions which can be reached by each arm, in the hypothesis that the movements take place with the arm in the condition of

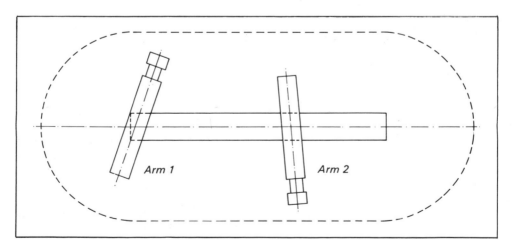

Fig. 5 Work area of two-armed robot

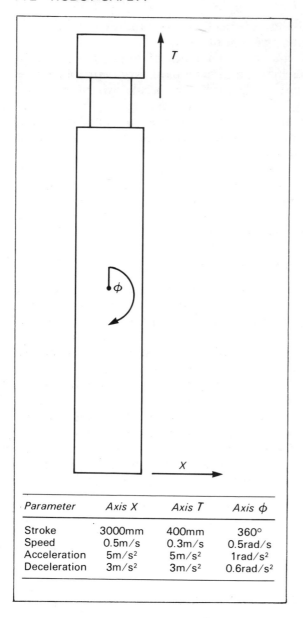

Fig. 6 Kinematic parameters of robot arm

Parameter	Axis X	Axis T	Axis φ
Stroke	3000mm	400mm	360°
Speed	0.5m/s	0.3m/s	0.5rad/s
Acceleration	5m/s²	5m/s²	1rad/s²
Deceleration	3m/s²	3m/s²	0.6rad/s²

maximum extension. This is not strictly true, but favours safety and noticeably simplifies the problem.

With reference to Fig. 8 and treating the three movements separately, we have:

$$X_{fA} = X_{iA} + v_A R + a_A R^2/2 + (v_A + a_A R)^2/2d_A$$

$$X_{fT} = X_{iT} + v_T R + a_T R^2/2 + (v_T + a_T R)^2/2d_T$$

$$\phi_f = \phi_i + \omega R + \dot{\omega} R^2/2 + (\omega + \dot{\omega} R)^2/2\dot{\omega}_d$$

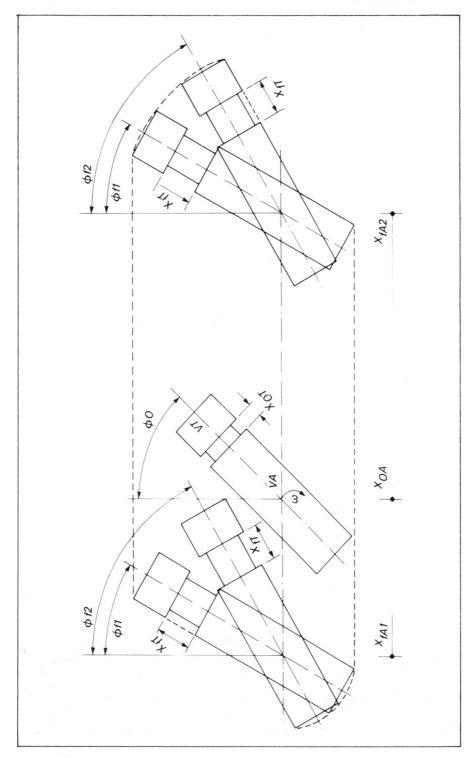

Fig. 7   Potential danger areas of one robot arm

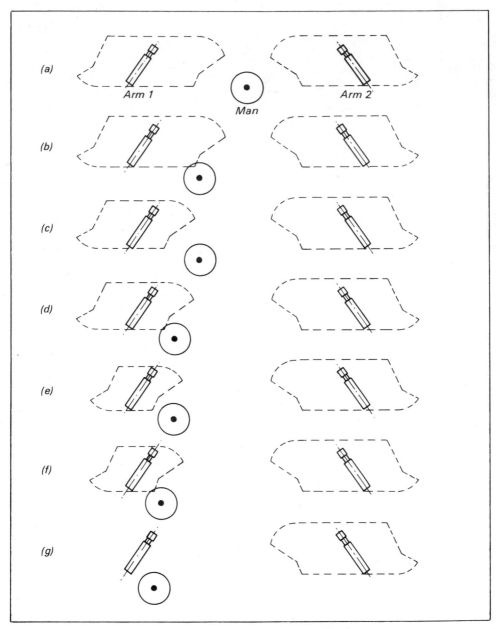

Fig. 8   Example of working system in the presence of a human invasion: (a) no inter-section between danger areas (normal scansion times guarantee safety); (b) intersection of danger areas of arm 1 and man (system recognises danger of collision between arm 1 and man); (c) safety system removes condition of danger by reducing scansion time of arm, consequently danger area of arm becomes smaller; (d) as a result of movements the danger areas of arm 1 and man intersect again (danger condition); (e) danger condition has been removed by reducing kinematic parameters of arm 1 (danger area of arm 1 has been further reduced); (f) due to further movements the danger areas of arm 1 and man intersect again (danger condition); and (g) danger area of arm 1 cannot be further reduced, the arm is stopped

where X_{iA} is the initial position of the axis of the arm along axis A, X_{iT} is the initial position of the head along axis T,ϕ_i is the initial angle formed by the axis of the arm with axis A, v_A is the speed of the arm along the axis A, v_T is the speed of the head along axis T, ω is the angular speed of rotation of the robot, a_A is the acceleration of the arm along axis A, a_T is the acceleration of the head along axis T, $\dot{\omega}$ is the angular acceleration of rotation of the robot, d_A is the deceleration of the arm along axis A, d_T is the deceleration of the head along axis T, ω_d is the angular deceleration of rotation of the robot, and R is the intervention time of the safety system.

Now, since the acceleration can have two maximum values of equal modulus and of opposite direction we have two extreme solutions for each movement. Ignoring the solution given by the acceleration along the T negative axis, we can derive the extreme positions occupied by the arm and hence determine the limits of the area of potential danger.

The possible presence of other objects in motion inside the area controlled leads to the definition of further areas of potential danger. If these areas intersect, we are in a condition of danger.

The removal of such a situation is possible by modifying the delay time of the safety system; where this is insufficient it is possible to modify the kinematic parameters of the machine. If the latter provision were still insufficient, it would be necessary to stop one or both of the arms. In the proposed example, the safety system could derive the information about the movement of the two arms from encoders applied to the arms themselves. The presence and movement of possible foreign bodies could be detected by a network of optical sensors or by microwave transducers.

Fig. 8 shows how the system works when invaded by a person. The first restricting of the area of potential danger of the robot is due to the reduction of the delay time of the safety system; the second is due to the limitation of the kinematic parameters of the robot, a further approach leads to the blocking of the arm of the robot, which cannot maintain the prescribed level of safety in any other way.

The algorithms determining the areas of potential danger and their intersections have been tested on a DEC PDP 11/34, and response times for the system of the order of 20–25ms were found. Taking into account the fact that these algorithms were implemented in FORTRAN, and that a noticeable diminution of computing time is possible by operating directly in ASSEMBLER, the hypothesis of response times of the order of 10ms seems to be confirmed.

Concluding remarks

The approach to safety problems illustrated above obviously has need of further investigation and development before its certain operative validity can be affirmed. On the other hand, the examination of the theoretical definition has not revealed any substantial difficulties of realisation. Having said this, it appears necessary at this point to carry out an experimental check with the purpose of determining the results which can actually be obtained via a computerised safety system. The experimentation should be developed, with reference to a well-defined case, in the following phases:

- Assessment of the performance which can be obtained from various types of transducers, possibly leading to the production of specific transducers.
- Emulation of the safety computer by a general purpose electronic calculator with the purpose of reducing the statement of the algorithms which could be written in evolved language.
- Definitive choice of microprocessor most suited to the realisation of the safety system and possible development of the specific hardware.
- Optimisation and implementation on the microprocessor of the algorithms for management of the safety system.
- Assessment of the results and comparison of the computerised safety system with traditional systems.

Design of Safety Systems for Human Protection

M. Linger
IVF (The Swedish Institute of Production Engineering Research), Sweden

In order to reduce the number of accidents in robot applications, IVF in Gothenburg is working on the design of safety systems for human protection. The three main areas of work being conducted are: analysing the requirements for safety systems for robots, developing safety systems based on a safety structure, and testing subsystems (sensors, control systems, power and brake systems). The aim of this paper is to inform on the design principles and some possible solutions for safety systems that allow people to have access to the robot's working space without interrupting production and safety systems for robot programmers.

In Sweden, from 1979 to 1981, 29 accidents in robot applications were reported[1]. This is approximately one accident per 100 robots a year. Half of the accidents occurred during the normal automatic cycle and the rest during maintenance, service, testing and adjustment. The most common reason for an accident was due to personnel repositioning incorrectly placed details. (For an up-to-date survey of robot accidents in Sweden, see page 49).

Following a fourteen day study, at a Swedish company, to find out the risk frequency in robot applications[2], it was found that 22 events occurred which could have been accidents and one accident occurred in which someone's finger was squeezed.

The main reasons for these accidents and the near accidents were:

- Human mistakes.
- Machine failure.
- Lack of safety system.
- The safety system was not 'production adapted' (i.e. safety system based on a sound knowledge about the process).

The production adaptation of a safety system is of the utmost importance. If its use causes too much trouble it is unlikely that the safety system will be used properly. A safety system must be production adapted and give protection in spite of machine failure or human mistakes.

A real production adapted safety system allows personnel to work within the robot's working area during the automatic cycle without any risk of injury due to machine failure or human mistakes. The robot only has to be stopped when someone gets 'dangerously' close; how close is determined by the robot's stopping time and distance and the type of safety system. Such systems are being developed and designed at IVF in Gothenburg. It is, however, important to remember that to have a safe working place, the safety system must be combined with education about how to use the machine correctly and information on the limitations of the safety system.

Safety system

A safety system consists of three equally important parts (Fig. 1):

- *Detection* of personnel in dangerous proximity to a robot.
- *Control system* which receives the alarm signal and transmits it to the power and brake system.
- *Power and brake system* for the robot's arm and gripper.

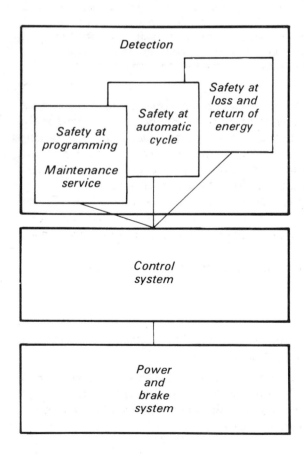

Fig. 1   Safety structure

Fig. 2   Mechanical guard

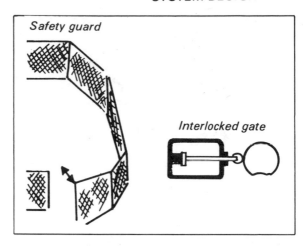

Safety guard

Interlocked gate

Detection systems

Mechanical guards. The usual way to build a safety system in robot applications is to use mechanical guards with interlocked gates (Fig. 2). The philosophy for such systems is that it should not be possible to run the process (robot) if somebody opens the gate and walks in. This can be a good solution during the automatic cycle if the production is working well. But today's robot applications do not allow wide tolerances and this causes many production stops.

We have already mentioned robot applications where positioning of details poses problems. To solve this, personnel may be allowed inside the mechanical guards to put the detail in the right place every cycle. This is typical of automatic systems. They are not made for people to be inside, but the easiest way to solve failures in design is to let somebody get inside to take care of the problems. Present day systems have many breakdowns and production stops and hence require production adapted safety systems making it easy to restart production.

Mechanical guards do not give any protection for people working inside the fences during programming, adjustment, maintenance and service. The mechanical guards themselves can cause squeezing risks between the robot and the fence if they are put too close to the robot. Personnel should be able to get outside the robot's working area if something happens.

A good alternative is to use lightbeams or lightcurtains (see page 199). This eliminates the squeezing risks between the fence and the robot. It is also easier to gain access to the robot. However, this also does not give any protection for personnel who have to work in the robot's working area. One way to reduce the risks is to let the robot run only at low speed. This is often possible during programming when one has to stand near the robot. However, low-speed operation is dangerous if, for example, the robot does not stop and if there is restricted space in which to stand beside the robot.

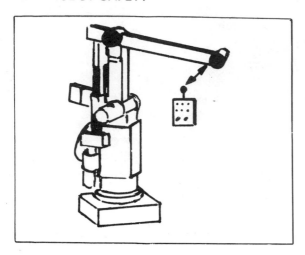

Fig. 3   Distance and direction to programming unit

Deadman's button. Some programming units are equipped with a 'deadman's button'. Safety then depends on the reaction speed of the programmer following dangerous robot movement. The philosophy is that he will release the button if he is going to be hit by the robot.

Distance and direction sensors. One idea is to have a distance and direction sensor on the programming unit (Fig. 3). If the robot is in dangerous proximity to the operator it will receive a stop signal. It would also be possible to allow only the slowest speed when they are close together. In this case the speed of the robot must be supervised.

The direction part of the unit is to ascertain whether or not the operator is turned towards the robot.

Safety sectors. Another way to design a safety system is to divide the dangerous area into sectors (Fig. 4). The philosophy here is that the robot should not be able to move into the sector in which a person stands. In this case, the position of the robot and the person has to be detected in a safe way. This can be done by using cams and switches to detect the position of the robot and sectors. This solution can give protection during automatic cycle, programming, maintenance, service and adjustments.

Sectors can be made using different types of sensor:

- *Contact mats* – this is a solution which can be easily implemented because of the availability of approved mat types for use in automatic systems (see page 205). They can be ordered in the shape which is suitable for the robot application (e.g. in sectors).

- *Passive infrared sensor* – this detects a sudden change in radiation in the infrared wavelength. In many applications, the infrared radiation (heat) from a human is greater than the radiation from the surroundings. If a person goes into the detection zone he will be detected by the sensor which

Fig. 4   Safety sectors

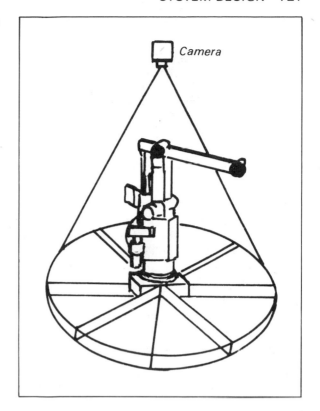

Camera

will issue a stop signal. To use this solution one has to ensure that there are no other moving objects having a human temperature (or greater), and corresponding size, in the detection zone and that there is not too much variation in reflections from lights.

- *Lightbeams/curtains*–the sectors can be implemented using lightbcams mounted beneath the working area. The distance between the beams must be close enough to even allow the detection of a human leg. Again, there are approved lightbeams and curtains commercially available.

- *Camera and image sensors*–this equipment is often used as a burglar alarm whereby the sensor detects a change in contrast. This could be developed into a safety system which detects the positions of the human and the robot. If they are too close together a stop signal is given to the robot. A camera and an image sensor can also be used to limit the working area of the robot and at the same time function as a fence around the working area.

Detection zones around the robot arm. The best way to design a safety system would be to have sensors around the robot arm (Fig. 5). If someone comes too close to the robot the stop signal is given.

At IPA in Stuttgart one such system has been developed. It is a soft electrical material which can be attached to the robot arm. If the robot touches something, the conductivity of the material will change and a stop signal is

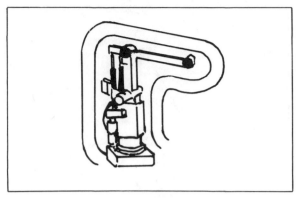

Fig. 5   Detection zones around the robot arm

given. This solution is a step towards a safe workplace for people who work in the robot's works area. However, there is still one more problem to solve before it can give complete protection for personnel, i.e. how to put sensors on the gripper. The robot can, for example, hold grinding machines and these can be more dangerous than the robot. Nevertheless, having soft material around the robot arm is still advantageous and would probably have prevented some accidents previously reported.

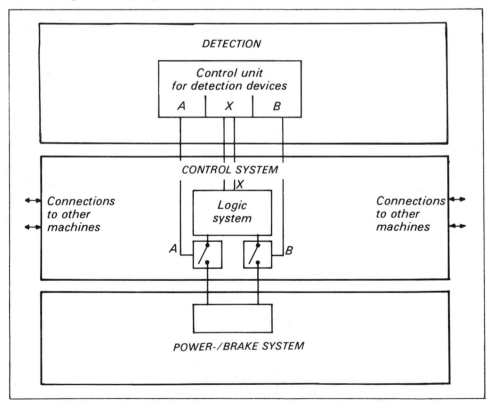

Fig. 6   Structure of safety signals: X are communication signals, A and B are safety signals

Control system

It is not enough to have just safe detection of personnel in dangerous proximity to the robot. The detection signal must also be connected to a reliable control system which processes the signals in a safe way. But what is a safe way? The preliminary demand is that one component failure should not cause a dangerous situation. One way to solve this is to have double stop and start signals. These are supervised so that if one safety signal fails it will be detected and the robot will stop. The probability of two signals failing at the same time is the square of the probability for one of the components ($10^6 \times 10^6 = 10^{12}$) failures per cycle).

The requirements for contactless detection systems in automatic press lines is that they should be doubled and supervised. The same requirements can be assumed for robot applications. The second requirement is to separate the safety system from the normal control system, since one disturbance with enough strength could 'knock out' two components at the same time.

Fig. 6 describes the principles. The detection system transmits two stop signals (A and B). If one signal fails it is detected and the robot is stopped. This fulfils the requirement that one component failure should not lead to a dangerous situation. The only safety element in the control system is the two connection points between the control system and power and brake system. Here, two signals are transmitted from the detection system in a certain amount of time after the detection of a person. This gives the maximum stopping time for the robot. The X signals are transmitted before signals A and B. This gives the robot the opportunity to make a controlled stop using its own logic system. If, however, there was a time delay in the logic system, the final stop would be made through the A and B signals.

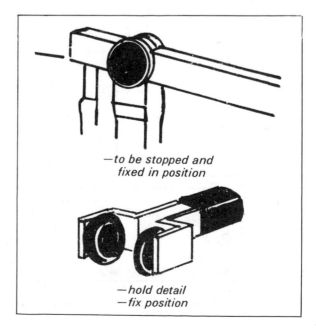

—to be stopped and
fixed in position

—hold detail
—fix position

Fig. 7 Axis of the robot and gripper

Using this solution there are undoubtedly possibilities to apply the safety systems on today's robots. It also gives unlimited possibilities for making changes in the control system without interfering with safety. As the robot is seldom working alone, it is therefore also important to ensure connections to other machines is made in a safe way.

Power and brake system

The power and brake system is concerned with both the robot arm and gripper (Fig. 7). Arm movement has to be stopped and fixed in position; the grippers must retain their holding force and be fixed in position – this must function in spite of energy loss.

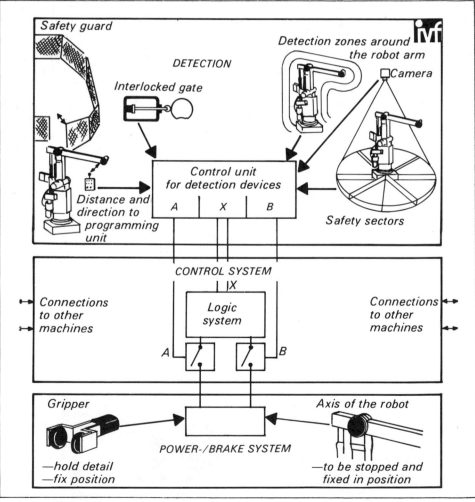

Fig. 8 Integration of three parts of the safety system, with their associated alternatives and possibilities

However, there are not many robots and grippers which fulfil these demands. Many grippers are pneumatically controlled and hence details would be dropped following energy loss. The problem is the same for pneumatic robots. Upon energy loss they will collapse and if a stop signal is given the already started movement cannot be stopped until it has reached the end position.

However, these problems are not impossible to solve. Power and brake systems are of a pneumatic, hydraulic, electrical or mechanical type. Presses and shears have such systems. Many accidents have occurred in these machines and therefore the safety demands are high. For manually operated presses with open dies, one of the demands is that one component failure should not lead to the press suddenly starting or not stopping if the operator activates the safety system. IVF were working with the type approval of presses and shears in Sweden between 1973 and 1981 and know that it is possible to fulfil these requirements. The same solutions can probably be adopted to the power and brake system of a robot, and IVF are presently studying this with different types of robot.

Finally, Fig. 8 summarises the three parts of the safety system with their associated alternatives and possibilities.

References

1. Backström, T. and Ringdahl, L.H. 1981. *A Statistical Study on Control Systems and Accidents at Work*. Occupational Accident Research Unit, Stockholm.
2. Tiefenbacher, F. 1982. *Occupational Safety with the Use of Industrial Robots*. Research Project, Swedish Work Environment Foundation (ASF), Stockholm.

Design for Safeguarding

J. P. Bellino
General Electric Company, USA

The safest robot installation is one that prevents personnel from entering the cell and projectiles from leaving. Since this is not practical in all cases, safeguards should be designed into systems. To date, the majority of robotic accidents have occurred during the programming/teaching mode or during the troubleshooting/maintenance mode. The cause of these accidents has been either equipment failure or human error. With this in mind, a design approach to help prevent these accidents from occurring is discussed.

The primary objective of safeguarding is to prevent injury to personnel and damage to equipment, particularly the programmer/teacher, the maintenance personnel, the operator, the equipment, and the workpiece. A minimum of the first three items needs to be addressed in all designs. Once the decision is made with regard to the objective of the safeguarding, the next step is to look at the overall robotic cell. Here, we first need to consider the layout of the system for such items as:

- Type of system: welding, material handling, assembly, etc.
- Methods of material transfer: conveyor, carts, positioners, etc.
- Ease of access for teaching, maintenance, material handling.
- Pinch points, guards, sharp edges, etc.
- Type of robot: gantry, SCARA, multiple arms, etc.
- Location of power drops, control cabinets.
- RFI and EMI sources.
- Manufacturers recommendation.

Safeguarding methods

At this point in the design, the basic methods of safeguarding need to be considered. The current technology offers a number of methods and devices that can be used either singly or in combination. Some of the most sophisticated and promising systems being developed utilise ultrasonics, microwave, infrared, and capacitance devices controlled through software that can detect the presence of humans. The more traditional designs of microswitches, light beams, pressure pads, etc. have been used for a number of years and form a good reliable base. It is likely that a combination of the above will form the basis for safeguarding systems:

When the basic approach has been decided and the hardware is being selected, the following items should be paramount in the ultimate selection:

- Hard to bypass.
- Simple to use.
- Fail safe.
- Reliable and highly immune to false triggering.
- Non-fragile.
- Hardened against electrical noise and industrial ambients.
- Ease of installation and repair.
- Cost effective.

Once the system has been defined, a risk analysis needs to be considered. The common modes that should be considered are: programming/teaching, troubleshooting/maintenance, and normal operation. When considering these modes, the analysis needs to be considered for all equipment functioning properly, then and at least a 'single' failure.

The robot installation

A typical robotic cell is shown in Fig. 1. Here, the basic components are robot, positioner, welder, and possibly a programmable controller. The primary function of the cell is welding with a human interface with the robot through a positioner. From a safety standpoint, we first need to look at areas of concern, then integrate them with the modes of operation.

First the cell is broken down into three discrete work envelopes:

- Work envelope of robot and end-effectors (Fig. 2(a)).
  Function   – area to perform required tasks
  Definition   – typically indicated by floor markings
  Safety concerns   – allow for access during teaching and maintenance

- Work envelope of positioner, tooling and parts (Fig. 2(b)).
  Function   – transition work to robot, and position for robot access
  Definition   – typically indicated by floor markings
  Safety concerns   – interlock such that personnel within reach will prevent positioner movement

- Work envelope outside robot and positioner envelopes, but within the perimeter boundary (Fig. 2(c)).
  Function   – keep projectiles in and personnel out
  Definition   – minimum 2m perimeter barrier with interlocked gates
  Safety concerns   – interlock such that intrusion shuts down system and projectiles are contained

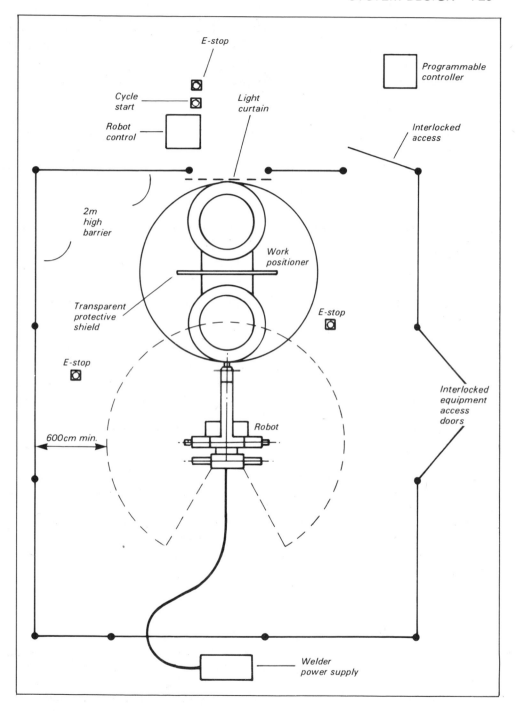

Fig. 1 Typical robotic cell

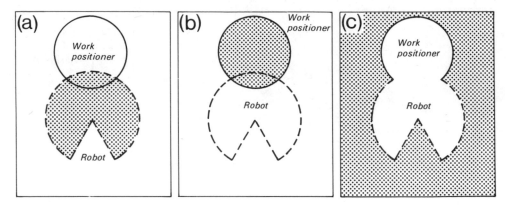

Fig. 2 Work envelopes: (a) of robot and end-effectors; (b) of positioner, tooling and parts; and (c) work envelope outside robot and positioner envelopes, but within the perimeter boundary

The teaching/programming mode of operation is now superimposed on each of the work envelopes. First, concerning the work envelope of the robot and how it relates to the positioner (Fig. 2(a)), as would normally be the case the positioner would be locked and the teaching carried out. Therefore, the primary concern is the working envelope of the robot.

Once the robot has been taught for one end of the positioner, the positioner work envelope is now interfaced with the robot (Fig. 2(b)). Here again, from the safety design standpoint, we can concentrate on one working envelope.

When the teaching has progressed to this stage (or sometimes sooner) we must look at both envelopes and the area outside (Fig. 2(c)). Again, a review of previous concerns.

Upon completion of the teaching/programming mode, a similar review for troubleshooting/maintenance, normal operation and other modes pertaining to the system can be made.

By breaking down the system into work envelopes, the modes of operation can be superimposed. Thus allowing an effective safety system to be developed in discrete elements. Once the design has been completed a final functional checklist can be reviewed (see Appendix to this paper).

Concluding remarks

Can absolute safety be achieved with robots? Most engineers will probably agree it cannot. Therefore, a blend of safety with efficiency of operation should be used in forming realistic goals. In most cases, as designs become more complex, the reliability of the overall system tends to decrease. With this in mind, the trend towards robots with the ability to detect the presence of humans will increase both the level of safety and reliability in the automatic mode. However, the safety of both the programmer/teacher and maintenance personnel must still be addressed.

References

1. *Safety Standards for Industrial Robots and Robot Systems,* Preliminary Draft – September 1984. Robotic Industries Association, Dearborn, MI, USA.
2. Schreiber, R. R. 1983. Robot safety: A shared responsibility. *Robotics Today,* October 1983: 61–65.

Appendix – Final functional checklist

Mechanical
- Does the peripheral barrier limit unauthorised access to the cell?
- Does the peripheral barrier allow for maximum reach of the robot, without causing a pinch point?
- Is the structural integrity of the barrier sufficient to stop any projectiles loosened from or released by the robot gripper?
- Does the system require remote material handling equipment with the robot such as conveyors, positioners, and shuttle systems; have the proper interfaces with the safety system been included?
- Are gears and pinch points covered with guards?
- Have all sharp corners and edges been eliminated or covered, especially on the grippers?
- Are the robot control panels located outside working envelopes and with an unrestricted view of the work areas?
- Has the robot been secured to its mounting surface to prevent accidental tipping or movement?
- Have all hoses been secured to prevent swalling in case of a ruptured line?
- Has adequate room been provided for personnel during teaching and maintenance?

Electrical
- Have all intrusion detectors been wired in series with emergency stop circuits on the inside and outside of the work envelope?
- Have adequate controls been installed to prevent a collision of movable equipment in the event of a software failure?
- Is an emergency stop switch readily available during robot teaching which will stop all relevant equipment?
- Have emergency stop switches been located both in the cell and outside the cell in readily accessible areas?
- Has a warning light been included in proximity to the working robot to indicate when robot movement can occur?
- Is the design of the gripper such that parts will not be dropped due to loss of power?
- Have pressure sensors been included to monitor hydraulic and pneumatic supplies and prevent the robot from operating with either too high or low pressure?

Procedures
- Has the design been researched to ensure that all federal, state, local codes and regulations have been complied with, including the equipment manufacturer's recommendations?
- Has the system been designed to require a minimum of two manual actions prior to restart?

- Has a system been put in place to ensure that all safety violations have been removed prior to a system start?
- Is the material being handled safely with regard to toxicity, flammability, radio-activity and, as required, by applicable regulatory agencies?

CAD – An Aid to Robot Safety

Y.F. Yong, BYG Systems Ltd, UK,
N.K. Taylor, Nottingham University, UK,
and M.C. Bonney, Loughborough University of Technology, UK

How CAD may be used as an aid to design safe robot systems is discussed. The paper is divided into three parts. The first examines how CAD could help with the design of safe robot systems and defines the area in which computer graphics can make an effective contribution. The second describes the GRASP package, a computer aided design system specifically constructed to aid the design of industrial robot systems. Thirdly, some special purpose additions to GRASP are described which illustrate some of the ways in which CAD could be developed specifically for robot safety evaluations.

Some of the safety problems of industrial robots may differ from those associated with conventional machinery. Unlike conventional machines which are basically operating in a stationary mode, the joints of robots are capable of moving within prescribed areas as defined by the manufacturer. This capability, whilst providing the industrial robot with its flexibility has also made all locations within the reach envelopes into potential risk zones. Many early robot applications have been in the areas where the work was hazardous. It would therefore be a paradox if in trying to make a job 'safer', robot systems themselves are not made as safe as is reasonably practicable. Safety should be an integral element of the design process and not an afterthought.

As robot technology develops, so will the complexity of its applications. Future installations are likely to use the industrial robot as an element which has to interact with other equipment within a complex system.

How can CAD help robot safety?

Computer aided design (CAD) can be described as the application of computers to design where the designer converses directly with the computer by using a graphic or non-graphic console in such a manner that his problem solving processes are highly responsive and essentially uninterrupted.

CAD does not by itself provide solutions, but helps the designer to provide solutions. It does not remove the onus from the designer. Employed sensibly, CAD can often provide improvements to the final design.

The increasing complexities of present and future robotic systems provide CAD with ideal opportunities for positive involvement. Interaction between the components of a system is quite often impossible to evaluate manually (i.e. through drawings in a two-dimensional static mode). CAD allows evaluations to be carried out in an interactive manner. The use of computer graphics allows ease of visualisation, particularly if a geometric modeller is available.

There can be many advantages associated with the use of CAD. It allows engineers to consider alternative options in design, and it also has the advantage of speed. Quite often it replaces the need to build expensive mock-ups. CAD provides flexibility for rapid design changes.

To apply CAD to the problems of safety design we need to establish the instances where robot/human contact takes place. Generally, these can be categorised into three main types:

- *Essential contact between man and robot.* Examples of this are programming in a teach mode and maintenance procedures, each of which is likely to require man and robot to share the same workplace.

- *Human error.* Within practical limits, robot workplaces need to be designed to prevent human error occurring in potentially dangerous situations. For example, appropriate guarding will, in general, prevent persons entering the workplace. Detection and reaction to the presence of a person in the workplace will be useful and the provision of carefully sited emergency stops is important.

- *Aberrant robot behaviour.* This may be caused by malfunctions of software or hardware occurring separately or jointly. These could take the form of mechanical, electrical or hydraulic failure or by software errors being present or arising.

It is in the first two categories of contact that CAD can contribute towards making robot systems safer. CAD can be used for this in the following ways:

- Programming robots off-line by CAD. This will reduce the level of essential contact as described above.
- Designing and planning the overall layout of robot installations. For example, CAD will enable guarding to be designed and can be used to identify potential trapping points. The interactions of the robot with other equipment within the installation can be simulated.
- CAD can allow for a more systematic approach towards safety. The reach contours of robots can be determined and a risk zone concept introduced into an installation. This is discussed in more detail later.

GRASP for designing industrial robot systems

GRASP is a CAD system which uses computer graphics to simulate robot installations. It was originally developed at the Department of Production Engineering and Production Management at Nottingham University, and currently, further developments are being carried out at BYG Systems Ltd, Nottingham.

Using GRASP, the design of FMS/robot cells can be optimised, thus saving time and money. It is a fully integrated modelling and simulation package commercially available in a variety of configurations. The system has been successfully used to evaluate a number of robot installations. Typical outputs from a GRASP simulation are shown in Fig. 1. The main features of GRASP are:

- Three-dimensional solid modeller giving hidden lines capability.
- Powerful display and manipulation facilities.
- Generalised kinematic robot modeller allows the user to model his own robots.
- Sophisticated robot control allowing creation, storage and replay of robot programs – including robot diagnostics.
- Path control and automatic interference detection.
- Multiple robot simulation and synchronisation.
- Simulation of non-robotic equipment and peripheral devices, e.g. conveyors, AGVs.
- Estimation of cycle times.
- Generation of data for creating off-line robot programs.

Some of the major benefits that can be obtained from using a package such as GRASP are:

- Optimise design of robot cell or FMS at graphics terminal.
- Compare and evaluate different robots for a particular task.
- Identify problem areas during planning and design.
- Reduce overall project time from design to installation.
- Reduce possibility of mistakes and hence increase cost savings.
- Aids visualisation through computer animation.

Programming robots off-line by CAD

Presently, the method of programming robots is by the teach mode, either by using a remote control pendant or a teach arm. From the safety point of view, although unavoidable this situation is clearly unsatisfactory. For many types of robot applications (e.g. arc welding), the operator has to work extremely closely during programming in order to obtain a satisfactory result. Even with the necessary precautions taken, such as reducing the speed, there will still exist potential dangers with close working. The answer must lie with reducing the essential man/robot contact.

The advances made in robot technology have meant that there are now control systems which are capable of taking off-line programs. It is technically

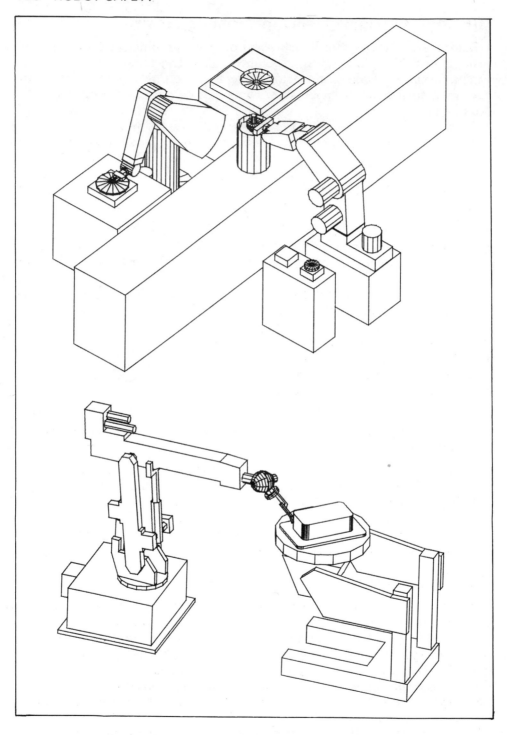

Fig. 1 Typical outputs from a GRASP simulation

possible to develop robot sequences off-line on a CAD system. These programs can then be loaded onto the robot controller and 'fine tuned') before being used operationally. The implications of such developments on safety are that the essential contact time for programming will be substantially reduced to that required for fine tuning.

Research and development work at Nottingham[1] and Loughborough Universities has demonstrated the viability of off-line programming. Although there are still many problems to be ironed out, commercial systems are beginning to appear on the market. Off-line programming provides the industrial robot with additional flexibility; it keeps robots operational whilst new programs are developed on CAD systems.

Some adaptations of GRASP for safety design

Available statistical evidence has revealed that the greater number of industrial robot accidents occur outside its normal running mode. In his Japanese survey (see page 23) Sugimoto found this figure to be as high as 90%, and Carlsson (see page 49) confirms that many accidents occur when the operator is inside the barrier surrounding the robot's area of operation to rectify faults or for cleaning.

The proper integration of the robot with other associated items of machinery is thus important. Due consideration on operator/robot separation during normal running is also important. GRASP can help identify areas of potential risk and thus provide the engineer with a powerful tool for improving his system and workplace design.

Planning the overall layout of an installation with GRASP is a relatively straightforward task. Models are built up from a combination of primitive shapes (e.g. cuboid, prisms, cylinders) and once these models have been input into the system, the user is provided with great flexibility. A comprehensive workplace/view menu allows the user to drag, reorientate and reposition models within the workplace – evaluating interactions between components within an installation. This evaluation process goes on until the user is satisfied with the layout. On CAD systems, objects can be moved with the 'stroke of a lightpen', so to speak.

Linked to the overall layout of an installation will be the guarding requirements. It is one of the essential safety features which require careful consideration. Guarding is necessary to prevent entry by unauthorised personnel and perhaps also to prevent any accidental escape of projectiles during a gripper malfunction. The determination of the height and layout of perimeter guards is thus important. In GRASP, this is assisted by the 'sling algorithm' which simulates the release of a projectile from a moving robot arm. Fig. 2 shows the front and plan views of the trajectory of a released object.

Trapping points

The ability to introduce a man model such as SAMMIE[2] on to the workplace simulation provides many advantages to the designer. Man models can be used

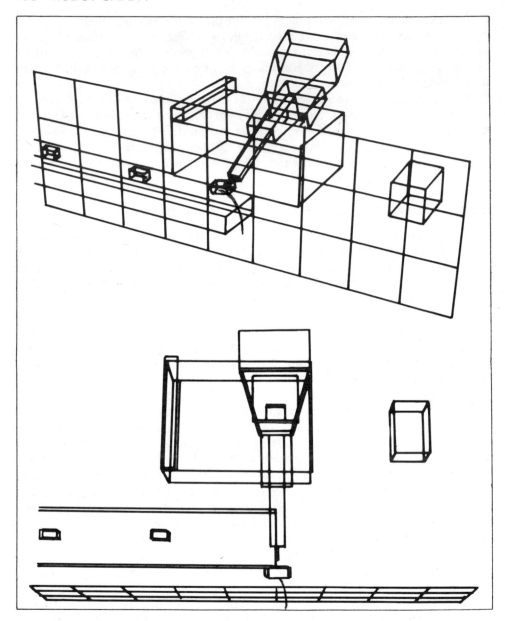

Fig. 2 Front and plan views depicting the trajectory of a released object

to simulate posture positions required for maintenance programming. They allow the identification of potential trapping points within an installation.

Experience by robot users has shown that the robot is sometimes capable of aberrant behaviour. Such behaviour is particularly dangerous if it should occur in the presence of people. In view of the numerous items of equipment which exist within an installation, there are bound to be many areas which can

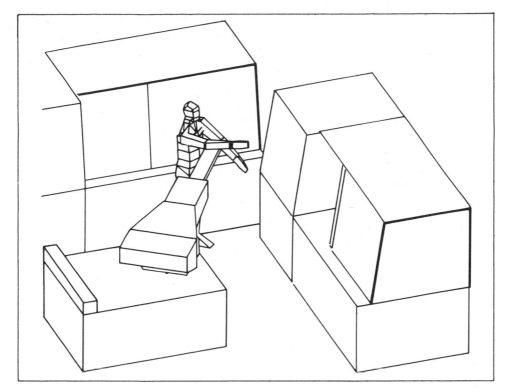

Fig. 3   An example of trapping in a cell

become potential trapping points. It is therefore essential to eliminate these trapping points by design; or if that is not possible, identification will allow safeguards (e.g. emergency stops) to be designed in. An example of trapping within a simple robot cell is shown in Fig. 3. CAD can assist in the detection of these potentially dangerous areas.

Operating zones and maximum reach envelopes

It is possible to introduce a risk zone concept into the operational area of a robot installation. Not all areas within the enclosure need necessarily carry the same risk weighting. Categorising areas into zones with varying risk levels may help decide on the action required to make an installation safe.

For the robot, we introduce two terms, the operating zone (OZ) and the maximum reach envelope (MRE) to assist in zone definition. MRE, as the name implies, can be taken to be the area described by the robot from a fixed position. OZ, on the other hand, is the area described by the robot in performing its task as required. These areas can increase if the robot is not stationary, e.g. if it is mounted on a moving track.

As an illustration, three zones could be defined in descending order of risk – red, orange and yellow. The OZ of a robot being of the highest risk becomes the red zone. The area covered by the MRE but outside the OZ

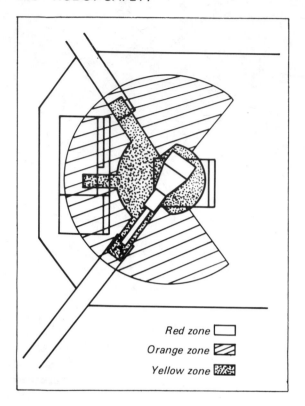

Fig. 4 An example of risk zones in a cell

Red zone □
Orange zone ▨
Yellow zone ▨

becomes orange. All other areas within the perimeter guarding of the installation will then be designated yellow.

In the GRASP CAD system, the definition of these zones can be achieved. Special purpose software could be developed to map these contours automatically. An example of a simple zone definition is illustrated in Fig. 4.

Acknowledgements

The authors wish to acknowledge the support provided by the Science and Engineering Research Council in funding the GRASP project. They also wish to acknowledge their former colleagues in the GRASP project. GRASP is now marketed by BYG Systems Ltd, Nottingham.

References

1. Yong, Y.F., Gleave, J.A., Green, J.L. and Bonney, M.C. 1985. Off-line programming of robots. In, *Industrial Handbook of Robotics,* Chap. 4.3. John Wiley, London.
2. Bonney, M.C., Blunsdon, C.A., Case, K. and Porter, J.M. 1979. Man-machine interaction in work systems. *Int. J. Production Research,* 17(6):619–629.

Safety Computer Design and Implementation

R.D. Kilmer, H.G. McCain, M. Juberts and S.A. Legowik
National Bureau of Standards, USA

There are many different aspects of safety to consider when utilising a robot in an industrial application. In general, however, these can be categorised into the areas of personnel safety and equipment safety. This paper addresses the later category and presents one approach of providing equipment safety through the use of an auxiliary computer to monitor operations in the workstation. Such a computer system can be used to check robot operations during programming, automatic cycling, and debugging and repair to prevent unwanted conditions from occurring. The basic concepts, design and implementation of such an auxiliary computer on a robot operating in a machine workstation are described.

Introduction

The problem of maintaining equipment safety in a workstation equipped with an industrial robot is a complex one to solve. Unlike other types of automated devices, a robot is not constrained in terms of what path in space it can travel or the velocity at which it can operate. This problem is not a severe one at present because most robots operate using pretaught programs which define the path through which they can travel. The velocity is also a predefined parameter and most robot controllers have internal checks to limit the velocities to some maximum value. However, as robotic systems become more sophisticated in terms of using sensory feedback for real-time control of the robot path, the problem of equipment safety will increase. Also, most robotic systems are relatively new devices. As a result, there is little or no statistical information regarding failures or what preventative measures might be taken. The only approach that can be taken is to examine the components which might be suspected of needing maintenance and to repair others after a failure has occurred. Thus, maintaining equipment safety will undoubtedly become more critical unless preventative measures are taken to minimise such problems.

One approach of addressing this problem is to use an auxiliary safety computer to monitor robot operations. The purpose of the safety computer is to prevent the robot from damaging itself or any of the equipment and sensors mounted on, and around the robot, in the event of a hardware malfunction, software bug, or operator error. To do this, the safety computer monitors a number of status and sensor inputs from the controller, the robot, and other sources. If the information that the safety computer monitors does not satisfy certain conditions, the safety computer stops the robot and notifies the operator.

Safety computer features

The safety computer, referred to as the Watchdog Safety Computer (WDSC), is a standalone microcomputer system used to monitor robot-related operations in the workstation. The purpose of this system is to detect operations which are outside the range of normal conditions and to stop the robot before a collision or any damage to the robot can occur. The initial implementation of this computer system is on a Cincinnati Milacron T[3] robot installed in the Automated Manufacturing Research Facility (AMRF) at the National Bureau of Standards[1]. In this application, the WDSC is used to monitor the individual joint and tool point motions of the T[3] robot and various status signals from the hydraulic pump unit and the T[3] controller. The WDSC measures the amount of rotation from the known 'home' position, the rotational velocity and the rotational acceleration of each of the six robot joints. It compares these measurements with a set of maximum values, which are operator selectable, and if the maximum values are exceeded, halts the robot. The WDSC also performs a similar set of comparisons for the tool point velocity and acceleration. The hydraulic pump unit and T[3] robot controller status checks are simple go/no-go tests. Fig. 1 shows a block diagram of the implementation of the WDSC on the T[3].

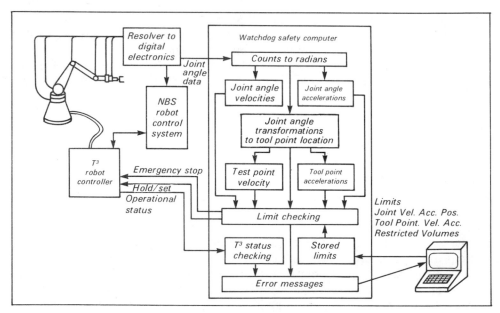

Fig. 1   Watchdog safety computer implementation on the T[3]

These maximum value and status checks are also made by the T^3 controller, so that, in this case, the WDSC is a redundant system. One difference is that the individual joint limits are operator selectable in the WDSC, which is an added convenience during system testing. The redundancy actually goes one level lower in that another resolver was installed on each joint so that joint position data could be obtained independent of the existing T^3 controller. In addition to the actual resolver installation, a resolver electronics system was designed and fabricated which provides joint position data for the six joints for up to three separate user systems. At present only two systems, the WDSC and the NBS robot control system[2], obtain T^3 joint position data from this resolver electronics system.

Also incorporated in the WDSC is a forbidden volume check. These volumes are regions in the work volume, such as a machine tool, an operator's workstation, or the floor, into which the tool point of the robot must not enter. The volumes are formed by representing the surfaces of the object with from one to six planes. These planes are programmed into the WDSC by moving the tool point of the robot to three points near the surface of each plane. At each of these points, the WDSC computes the tool point location in world coordinates and from these, the location and orientation of the plane in space. For each plane, a safety margin is assigned which accounts for the distance required to stop the robot. If the tool point of the robot attempts to enter one of these forbidden volumes, the WDSC halts robot motion. In this case, the WDSC is not a redundant system since the T^3 controller does not provide such forbidden volume checks.

As presently configured, the WDSC checks are performed only when the T^3 controller is in the auto mode. This covers all automatic operations when the robot is performing a taught program and for all conditions when the robot is being controlled by the NBS robot control system. The WDSC, although operating, does not halt robot motions during manual and teach operations due to limitations in the T^3 controller which only allow it to request a hydraulics shutdown. Stopping the robot by disabling the hydraulics is undesirable in many instances because the arm may settle down on an object and cause more damage than if the operator is allowed to correct the situation.

T^3 robot description

The T^3 robot is installed in a horizontal machining workstation in the AMRF. In this workstation, the T^3 is primarily involved in material handling tasks (Figs. 2 and 3).

The T^3 is a six-axis servo-controlled robot powered by hydraulic actuators. It is manufactured by the Cincinnati Milacron Company located in Cincinnati, Ohio[3][4]. All of the actuators, except the one on the elbow joint, are rotational actuators. The actuator on the elbow is a linear piston type. All of the joints, including the elbow, are rotational and give the T^3 six degrees of freedom. The actuators support and move the mass of the T^3 when it is in operation and provide a rated lifting capacity of 200kg.

When the hydraulic power to the T^3 is off, the upper arm of the robot rests on

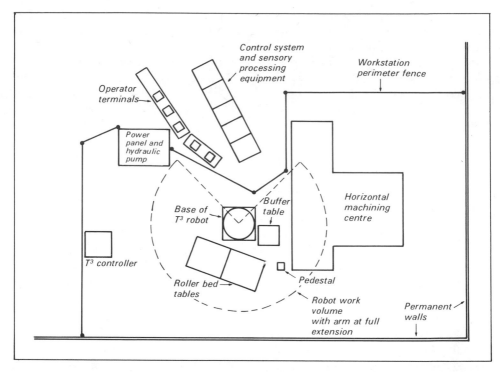

Fig. 2 Layout of the horizontal machining workstation in the AMRF

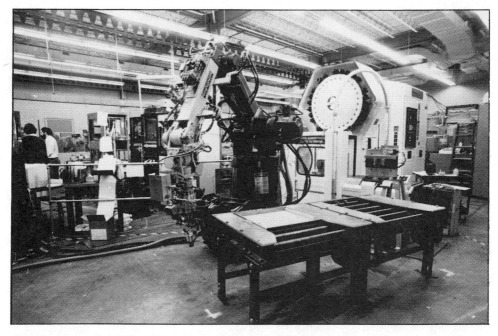

Fig. 3   View of the T³ robot as installed in the AMRF

a 'shot pin' at the shoulder joint. The shot pin extends automatically when the hydraulic power is off, preventing the shoulder joint from settling past a point approximately 45° above horizontal. If, however, the T³ is already below the shot pin, there is nothing to prevent the robot from settling down against the floor. The T³ could cause a great deal of damage if it settled on the end-effector or any of the equipment around the robot.

There are two control inputs to the T³ controller which the safety computer can use to stop the robot: emergency stop and hold/set. Both of these signals were intended to be used as manual shut-off switches by the robot operator. A number of normally closed switches are connected in series between the controller's ac source and the control inputs. When the circuit is broken by one of the switches being pressed, the T³ controller performs the requested function. The emergency stop input turns off the hydraulic power to the T³ and is active any time that the hydraulics are enabled. The hold/set input causes the T³ to stop its motion without shutting off the hydraulics. The motion can be resumed by issuing a signal on the hold/clear input. However, the hold/set input only works when the T³ controller is running in the auto mode. A hold/set will not stop the robot when it is being run in manual or teach mode.

One of the major research topics being pursued in the AMRF is the development of a generic robot controller. To that end, the NBS Robot Control Systems (RCS) is being used to run the T³. This system controls the T³ robot by sending commands to the T³ controller using an external communication link. This link is supported in the T³ system software by a program called Dynamic External Path Control (DEPC). DEPC allows the T³ to be controlled via a serial communication link with an external computer, in this case, the NBS RCS. (This feature is functional only when the T³ controller is in the auto mode.) In the future, the T³ will be run exclusively through this control link. The NBS RCS will provide the high-level control, and only the low-level servo control will be left to the T³ controller. As a result, the T³ will be running in auto mode most of the time, except for brief periods of time when the robot is being powered up or shut down, and thus can be stopped using a hold/set.

Safety computer hardware configuration

The safety computer is a multibus compatible microcomputer system. This system, shown in Fig. 4, consists of a multibus 9-slot chassis equipped with a single-board computer using an 8086 CPU, a parallel I/O board with three 24-line parallel ports, and two EEPROM memory boards. All of these boards are commercially available and do not require customising prior to use in the safety computer.

As shown in Fig. 4, the operator terminal is connected to the system through a serial port on the single-board computer. The joint position data from the resolver electronics are brought into the system through a parallel port also located on the single-board computer. The T³ status data and the hold/set and emergency stop output commands are transmitted through an optical isolator

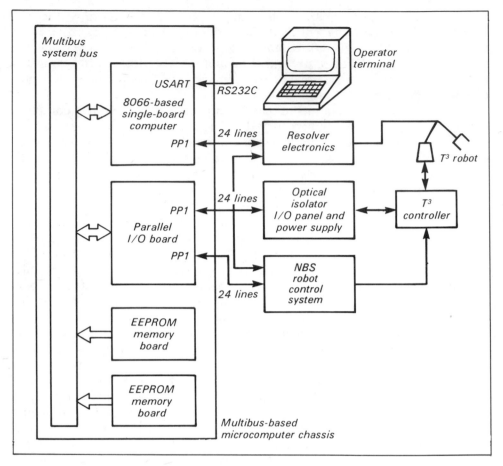

Fig. 4 Watchdog safety computer hardware configuration

I/O panel to one of the ports on the parallel I/O board. The other block shown in this diagram is the NBS RCS. The hardware and the majority of the software for this interface have been implemented, but have not been completely tested and debugged at this time.

The safety computer is located in its own chassis so that it is isolated from other systems controlling the robot. In this configuration, the safety computer has sole control of the system bus, and thus avoids the problem of multiprocessors using the same bus. This will increase system reliability and eliminate the possibility of the bus being tied up by another processor. If it shared the bus with the NBS RCS, for example, a malfunction in the RCS could quite easily cause the safety computer to malfunction as well.

As shown in Fig. 4, the safety computer is interfaced to the T³ controller through an optical isolator I/O panel. The safety computer receives a number of status indicators from the T³ controller through this interface. This is also used to send the hold/set and emergency stop signals to the T³ controller.

The T³ robot uses two different logic voltages, 120 V ac and 24 V dc, neither of which is compatible with the TTL logic used by the safety computer. The optical isolator I/O panel is used to convert between the robot's logic voltages and the safety computer's TTL logic. This panel has a total of 24 I/O lines which are configured (ac or dc voltage and input or output) by selecting appropriate driving or terminating modules. The modules convert from ac to dc to TTL, and vice versa. The interface between this panel and the safety computer is through the parallel I/O board. Both input data and output commands are transmitted over this interface.

Safety computer software structure

The general structure of the safety computer software is illustrated in Fig. 5. This figure shows the major components of the safety computer software structure and the interactions between these components.

The programming language used in developing software for the safety computer is FORTH. This software is divided into two major sections: the watchdog task and the operator task. The watchdog task is responsible for monitoring robot operations and stopping the robot to prevent any damage. The operator task provides an interface that allows the user to view the status of operations and alter the parameters of the safety computer. The two tasks timeshare on the same CPU[5].

The operator task runs the FORTH operating system and can be used to run programs that let the user examine or change parameters and to edit the watchdog program. Fig. 6 shows the user terminal display for editing the velocity, acceleration and stopping threshold limits. The operator can select the limits by specifying a percentage of the maximum for each parameter shown in this display. This value is the percentage of a maximum value (either velocity, acceleration, or stopping threshold) that was determined empirically

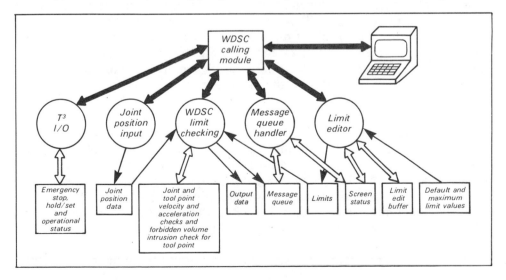

Fig. 5 General software structure of the watchdog safety computer

	Percentage of Maximum		
Joint	Velocity Limit	Acceleration Limit	Stopping Threshold
BASE	50	50	50
SHOULDER	50	50	50
ELBOW	50	50	50
PITCH	50	50	50
YAW	50	50	50
ROLL	50	50	50
Tool-Point	50	50	50

^ E – Enter a new value
^ D – Reset entry to default
^ R – Reset table to default
^ Q – Quit the editor
Use cursor control keys to select entry.

Fig. 6 User terminal display for editing limits

and is embedded in the software for checking these limits. If the operator does not select new limits, the default values of 50% are used. Fig. 7 shows the display for examining the position, velocity and acceleration values for the T^3 robot.

The operator task handles all communications with the safety computer's terminal, and hence the user. The watchdog program is transparent to the user in the operator task except when the watchdog program sends messages to the user via the operator task. The terminal is slow relative to the watchdog program; it is only able to print out about one character every millisecond at 9600 baud. Thus, messages generated by the watchdog program are stored in a message queue until they can be printed out. When the watchdog program has put a message, or messages, into the message queue, it instructs the operator task to run a program that will list out the contents of the message queue at the terminal for the user to see.

The watchdog program, running in the watchdog task, is activated every 50 ms by a timer interrupt. This interrupt is a signal to the watchdog program to check the status of the T^3. The program examines the data it receives and decides whether the robot is operating normally. If the data violate any of the robot's operational constraints, the watchdog program commands the T^3 to stop. The watchdog program instructs the operator task to print out messages indicating the status of the watchdog program and any error conditions that it has detected.

The watchdog program is broken down into three functional blocks: input, computations, and output. Each of these functional blocks decomposes into a number of more specific tasks. The input block reads joint data from the

Safety Computer Status Display

Joint	Angle radians	Velocity rad/sec	Acceleration rad/sec/sec
BASE	-0.7011	0.0111	-3.5166
SHOULDER	-0.6991	0.0205	2.2592
ELBOW	1.4293	0.1219	-1.5345
PITCH	0.0113	-0.0201	-0.3987
YAW	-0.0153	0.0402	17.3720
ROLL	-0.0428	-0.0229	2.1544

Tool Point	Position inches	Velocity in/sec	Acceleration in/sec/sec
X	69.9104	-5.4277	199.4445
Y	-59.7125	4.0924	194.7723
Z	-29.8131	-8.7384	-96.2600
Magnitude		11.0710	294.9244

Robot Operating Mode: MANUAL
Safety Computer Mode: MANUAL/AUTO
Message Queue Status: EMPTY
Press (HOME) to quit.

Fig. 7   User terminal display for examining the motion of the T³

resolver electronics and status data from the T³ controller. The computation block checks for timing and I/O errors, converts the data from the resolvers into joint angles, computes the position, velocity and acceleration of the robot, and checks for discrepancies in the robot's motion. The output block sends commands to the T³ controller (hold set and emergency stop), and sends error and status messages to the user via a message queue. A top-level flow diagram of the watchdog program is shown in Fig. 8.

In order to explain how the watchdog program acquires its data while allowing the user to interact with the operator task, an understanding of the multitasking environment of FORTH is required. FORTH uses a round robin multitasking scheme whereby all of the tasks are linked together in a circular list. The CPU passes control around this linked list processing each task in turn. Each task is in one of two states, awake or asleep. When a task is asleep it does not require any servicing by the CPU, and control is passed on immediately to the next task in the round robin. If the task is awake, the program in the task resumes execution where it left off. The CPU continues executing the program until the program releases the CPU to return to the multitasking round robin. The CPU can be released explicitly by the program or implicitly by certain FORTH words. The words primarily responsible for releasing the CPU are those concerned with terminal I/O. Whenever a character, or string of characters, is sent or received from the terminal, the task

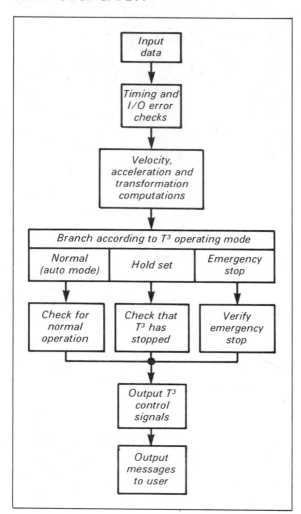

Fig. 8 Block diagram of the watchdog program in the WDSC

is put to sleep and the CPU is released to service other tasks in the round robin. While the task is asleep, an interrupt routine accepts or transmits data to the terminal. When the transfer is complete, the interrupt wakes the task so that it can resume operation when its turn comes again.

The CPU alternates back and forth between the watchdog and operator tasks. The operator releases the CPU when it is waiting for terminal input or when it is in the process of writing out to the terminal. The watchdog task releases the CPU when it has no new data to process. New data are read in from the resolvers every 50 ms by a routine triggered by a timer interrupt. When the interrupt routine has read in the resolver data, it increments the 'new-data' flag. This indicates to the watchdog task that there are new data to be processed, and that the next time control of the CPU is passed to the watchdog task it should execute the watchdog program. When the watchdog program has been activated by the interrupt, it collects all the other data necessary to make its decisions. In the present version of the safety computer, these are the T^3 status data.

When all of the data has been collected and the computations completed, the watchdog program decides whether to stop the robot based on the current state of the safety computer and the operating conditions of the robot. The operating conditions of the T^3 are given by the T^3 controller status signals and the position, velocity, and acceleration values calculated from the resolver data.

The state of the safety computer is dependent upon previous decisions and actions and determines how the watchdog program will respond to the current input data. The five states of the safety computer are:

- Manual/auto – the normal operating mode of the safety computer when the watchdog program has not detected any error conditions.
- Hold/set – an error has been detected and the safety computer has issued a hold/set; the hold/set line is still active.
- Hold/set – an error has been detected and the watchdog program has issued a hold/set; the hold/set line is inactive (the hold/set signal is momentary).
- Emergency stop – an emergency stop has been issued, because the T^3 failed to stop after a hold/set was issued; the emergency stop line is still active.
- Emergency stop idle – an emergency stop has been issued, because the T^3 failed to stop after a hold/set was issued. The emergency stop line is inactive (the emergency stop signal is momentary).

In each of these five operating modes, the watchdog program expects the data to meet certain requirements. The differentiation between hold/set and hold/set idle, and emergency stop and emergency stop idle, is the state of the output line. Both the hold/set and the emergency stop should be momentary signals on the control lines. While in hold/set or emergency stop mode, the signal line is being pulsed active. During the idle modes the signal is inactive. Thus, hold/set goes to hold/set idle, and emergency stop goes to emergency stop idle after briefly activating the respective control line. The safety computer stays in the idle state until something (for example, a hold/clear or a manual reset of the safety computer) causes the operating mode of the safety computer to change again.

When the T^3 is operating normally, the safety computer is in the manual/auto mode, and the following are checked:

- Joint velocities below maximum.
- Joint accelerations below maximum.
- Joint position within range.
- Tool point velocity below maximum.
- Tool point acceleration below maximum.
- Tool point outside all forbidden volumes.

If any of these constraints is violated, the safety computer will issue a hold/set. Each of these error checks can be individually enabled or disabled by the user appropriately setting or resetting the corresponding enable flag. The velocities and accelerations are checked by comparing the current value to the respective maximum allowable limit. Exceeding this limit is considered an error condition. Similarly, joint position is compared to a range of allowable values. Moving the joint outside this range is considered an error condition. The current tool point location is compared to the list of forbidden volumes. Moving the tool point inside the safety margin of any of the planes which make up these volumes is considered an error condition.

When a hold/set is issued, the safety computer enters the hold/set operating mode and checks to see if the T^3 has actually stopped. Since the T^3 is very large and the response time of the safety computer is finite, the robot will not come to rest immediately. To take this into account, the safety computer uses a stopping threshold to decide if the robot has responded to a hold/set command. The stopping threshold specifies how far the T^3 is allowed to travel after a hold/set has been issued. If the T^3 travels further than this stopping threshold, the safety computer assumes that the hold/set did not work and an emergency stop is issued. This check can be disabled by resetting the corresponding enable flag, 'e-stop-en'. This may be desirable during certain types of testing when an operator is present to manually hit the emergency stop switch if necessary. However, in most cases this will not be necessary because the stopping thresholds were empirically chosen so that there would be no false triggering.

In the emergency stop mode, the hydraulic power to the T^3 is shut off. As a result, the robot will slowly go limp, settling down on whatever is below it. With no power to the T^3 there is no control, and therefore, no controlled motion. The only check for emergency stop is to examine the appropriate status line to see if the controller is responding properly. If it is not responding, there is no action that can be taken on the part of the safety computer, aside from notifying the operator.

Resolvers

The safety computer's main function is to monitor the motion of the T^3 robot. This is done by monitoring the position of the robot with six resolvers, which are mounted in the robot's position analogue units (PAUs). Each PAU contains a resolver and tachometer which provide joint position and velocity feedback to the T^2 controller. An additional resolver was added to each PAU to provide an independent source of joint position data for the safety computer. These joint angles are used to compute the various motion parameters checked by the safety computer. Each resolver returns an analogue signal which indicates the angular position of the resolver and thus, the position of the robot joint. This analogue signal is converted to digital form by an R/D (resolver to digital) converter in the resolver electronics package designed and built at NBS. The R/D converter generates a 14-bit number which is representative of the resolver angle. A block diagram representation of the resolver electronics is shown in Fig. 9.

Due to the gearing in the PAUs, the resolvers go through a number of revolutions while the joint travels through its range of rotation. The R/D converters only provide the angle of the resolver. To obtain the angle of the joint, the number of revolutions of the resolvers must be taken into account. This is done by the resolver electronics package, which has a 3-bit revolution counter for each resolver and adjusts the revolution counter approximately when the R/D converter output rolls over. The three bits are combined with the 14-bit R/D output to produce a 17-bit result. In order for the resolver electronics to provide a consistent result, the revolution counters must be initialised every time the resolver electronics are powered up. This is done by moving the robot to a known 'home' location and then setting each revolution counter to the correct revolution count. The revolution count at the home position is chosen so that the 17-bit resolver output will vary continuously over the range of the joint (i.e. so that it will not roll over in the middle of the joint's travel). This is done manually from the front panel of the resolver electronics.

Both the safety computer and the NBS RCS obtain data from the resolver electronics. These data are provided on request on a first-come first-serve basis. Since there is only one set of six R/D converters, the resolver electronics has circuitry that only allows one system to access the resolvers at a time. A system requests data by asserting its request line, and then waits for the resolver electronics to answer the request by asserting the acknowledge line. When the requesting system receives the acknowledge signal, it is free to read the

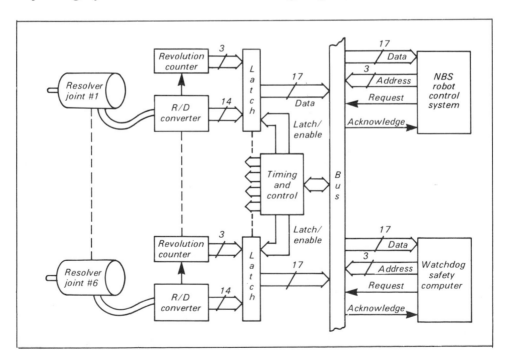

Fig. 9 Block diagram of the resolver electronics

resolvers. The resolvers are read, one at a time, by sending a 3-bit resolver address to each resolver channel and then reading the 17 bits of resolver data. All of the resolver data is latched at the moment the acknowledge is returned (forming a time-coherent set). Then the resolver address is used to select which latch is read. The acknowledge line stays active during the whole process of reading the resolver data and prevents any other system from accessing the resolver electronics. After a preset amount of time (approximately 1 ms at present), the acknowledge line goes inactive and other systems can request the resolver data. The safety computer monitors the amount of time spent waiting for the resolver electronics to return the acknowledge signal. If it does not respond within a reasonable amount of time, the resolver reading routine aborts and executes a hold/set on the robot.

The resolver reading routine is started by the 50 ms timer interrupt. In this way the interval between resolver samples is almost constant. The slight difference in intervals caused when both systems attempt to read the resolvers at the same time only results in a 2% error, which is acceptable for the velocity and acceleration calculations. The interrupt also sets the 'new-data' flag which notifies the watchdog program of the 50 ms interrupt. The watchdog program converts the digital resolver readings into the six joint angles in radians, which are then used in the forward transform, and the velocity and acceleration calculations. The conversion formula is:

$$A = (R * B / G) + O \tag{1}$$

where A is the joint angle, R is the integer resolver data, B is the radian bit weight, G is the gear ratio, and O is the resolver offset. The constants in the formula are chosen so that the joint angles produced are compatible with the forward transform. The radian bit weight is determined by the precision of the R/D converters, and is the same for all of the joints:

$$B = 2 \text{ pi} / 2 ** 14 = 0.000383495 \text{ rad/bit} \tag{2}$$

The gear ratio accounts for the gearing in the PAU between the robot joint and the resolver. The gear ratios for the six joints are given in Table 1. The gear ratios in this table have a negative sign if the positive direction of rotation of the resolver and the forward transform rotation conventions do not agree. This is discussed further in the next section. The resolver offset is used to make the joint angles agree with the conventions of the forward transform. The exact

Table 1 Gear ratios for the six joint resolvers

Joint name	Gear ratio
Base	8
Shoulder	-8
Elbow	-4
Pitch	4
Yaw	-4
Roll	4

values of the offsets depend on the physical orientation of the resolver with respect to the joint, and have to be recalculated any time this orientation is changed, e.g. when the PAU containing the resolver is removed or adjusted. The formula for calculating the resolver offset is:

$$O = H - (R * B / G) \tag{3}$$

where H is the known angle at which R was read. This is done at the robot's 'home' position. The T³ can be made to serve precisely to this 'home' position where these readings are made.

Tool point transformations

In addition to checking the motion of the individual joints, it is important to check the motion of the tool point, i.e. the point at the end of the robot gripper where an object is grasped. Since the motion of this point is a combination of the motion of all the joints, it is not measured directly but rather calculated. The formula used to compute the location of the tool point is called the forward transform. The forward transform is based on the physical geometry of the robot, including how it is jointed and how long the linkages are between the joints. The forward transform is a mapping from joint space to robot space. The joint space is defined by the joint angles of the robot, and the robot space is defined by the Cartesian coordinates of the tool point and the orientation given by three angles of rotation about the x, y and z axes.

The forward transform can be considered as a combination of simple transformations. Each translation and rotation can be thought of as a transformation from one coordinate system to another. Each linkage of the robot is thought of as having its own coordinate system, and then the transformation to change from one coordinate system to another is a simple translation and rotation. By combining these simple transformations, a composite transformation that goes from the tool point coordinate system to the robot coordinate system can be constructed.

The present version of the safety computer only needs the position of the tool point, not the orientation. This can be obtained by applying the transformation just described to the tool point. This is the origin in the tool coordinate system. In the safety computer, the individual linkage transformations are applied to the tool point to transform the coordinates of this point from the tool coordinate system to the robot coordinate system. Applying each of the rotations and translations in turn is more computationally efficient for a single point than constructing a transformation matrix from the product of the individual linkage transformations. This is not the case if several points need to be transformed or if orientation data are necessary.

In the interests of computational efficiency, the rotation and translation computations have been simplified as much as possible. Translations take the form of a simple addition, and rotations are specified around one of the three coordinate axes. The formula for rotating a point round an axis is given by:

$$U' = U * \cos (\text{angle}) - V * \sin (\text{angle}) \qquad (4)$$
$$V' = U * \sin (\text{angle}) + V * \cos (\text{angle}) \qquad (5)$$

Axis of rotation	U	V
x-axis	y	z
y-axis	z	x
z-axis	x	y

where the appropriate values are substituted for U and V depending upon the axis about which the point is being rotated. These simple transformations are sufficient to describe the articulation of the T^3. The sequence of transformations necessary to transform a point from tool coordinates to robot coordinates is given in Table 2. The forward transform is written assuming that all the angles are zero when the T^3 is stretched-out parallel to the floor. In practice, this configuration is physically impossible for the robot to obtain due to the limited range of travel of the elbow joint.

The current version of the forward transform takes about 5ms to execute for one point. This computation is done for every cycle of the watchdog program to calculate the position of the tool point.

Velocity and acceleration calculations

The velocities and accelerations used by the safety computer are computed in a relatively straightforward fashion are currently calculated using the following first order linear interpolations:

$$V2 = (P2 - P1) / T \qquad (6)$$
$$A2 = (V2 - V1) / T \qquad (7)$$

where $P1$ and $P2$ are the previous and current positions, $V1$ and $V2$ are the previous and current velocities, $A2$ is the current acceleration, and T is the interval between samples, or 50 ms. The joint velocities and accelerations are signed scalar quantities, with the sign indicating the direction of rotational motion. The tool point velocity and acceleration are vector quantities specified by their components along the three coordinate axes. The computation for the

Table 2 Forward transform sequence tool to robot coordinates

Transformation	Description of argument
x translation	tool point to roll joint length
x rotation	roll joint angle
x translation	roll joint to yaw joint length
z rotation	yaw joint angle
x translation	yaw joint to pitch joint length
y rotation	pitch joint angle
x translation	pitch joint to elbow joint length
y rotation	elbow joint angle
x translation	elbow joint to shoulder joint length
y rotation	shoulder joint angle
z rotation	base joint angle

vector quantities is essentially identical to that for the six joints. The calculations are done in a scalar fashion for each of the three coordinates (x, y, z), so that:

$$Vx2 = (Px2 - Px1) \, / \, T \tag{8}$$
$$Ax2 = (Vx2 - Vx1) \, / \, T \tag{9}$$
$$Vy2 = (Py2 - Py1) \, / \, T \tag{10}$$
$$Ay2 = (Vy2 - Vy1) \, / \, T \tag{11}$$
$$Vz2 = (Pz2 - Pz1) \, / \, T \tag{12}$$
$$Az2 = (Vz2 - Vz1) \, / \, T \tag{13}$$

The magnitude of the tool point velocity and acceleration vectors are used for comparison purposes.

These simple formulae work well as long as the measurements have sufficient precision and are relatively noise free. Fortunately, the resolvers fit the criteria nicely since they have almost no noise and enough precision to pick up very small changes in position. If this was not the case, some type of averaging would have to be done to reduce the effects of noise in the joint position data.

Forbidden volume algorithms

The forbidden volume routine provides a method for the safety computer to impose restrictions on the positioning of the tool point. In a sense, it is the three-dimensional analogue of the joint limits. The forbidden volume routine allows the safety computer to prevent the robot from damaging itself by colliding with some object in its work space.

In essence, the forbidden volume routine uses a simple 'world model' to predict when the robot is in danger of colliding with an object. This model consists of a number of 'forbidden volumes' which correspond to real objects in the work space. When the robot is about to run into something, it will cross the surface boundary of the object's volume. The volume can be defined larger than the object so that the robot reaches the surface boundary of the forbidden volume before colliding with the object. By defining the forbidden volume to be larger than the actual object, a safety margin is provided to allow for the time it takes the safety computer to perform its calculations, detect an error, and output the appropriate control signal to the T[3], and for the robot to actually stop. This response time and the velocity of the robot will determine how far the robot will travel before being stopped, and thus, the necessary size of the safety margin. Since the distance travelled in this time is a function of the velocity of the robot, the approach taken in the safety computer is to make the safety margin for the forbidden volumes a function of velocity. Thus, the faster the T[3] approaches a forbidden volume, the larger the safety margin will be to account for the longer distance required to stop motion.

A forbidden volume is defined as a collection of planes, or more accurately, as a conjunction of half-spaces. A half-space is bounded by a plane. Every point on one side of the plane is 'inside' the half-space. Thus, the forbidden volume is the region that is inside every one of a collection of half-spaces. Using enough half-planes in this manner, any convex object can be formed. (An object is convex if all of its surface planes meet in an outward pointing edge, like those of a cube.) If the object is not convex, it can be defined by a number of smaller convex objects. The actual complexity of the volumes defined in this manner is limited by memory space and available computation time of the safety computer. The processing of the preliminary forbidden volumes, which includes seven objects and a total of 17 planes, takes the forbidden volume routine about 12 ms. This fits easily into the 50 ms time cycle.

The forbidden volumes are stored as a list of planes, given by their plane constants, and grouped into objects. Each plane is specified by the four plane constants in the equation:

$$Ax + By + Cz = D \tag{14}$$

where A, B, C and D are the plane constants, and x, y and z are the coordinates of points on the plane when they satisfy the equation.

The actual half-space formula is specified by the inequality:

$$Ax + By + Cz \leq D \tag{15}$$

Given a point X, Y, Z, the point lies in the half-space if it satisfies the half-space inequality given above. The forbidden volume routines substitute the robot's tool point into the half-space inequality. If it satisfies the equation, then the tool point is within the half-plane. If it satisfies the equation for all of the object's half-spaces, it is within the forbidden volume.

The half-spaces for the objects are generated using four points. Three points are sufficient to define the plane, and the fourth point is used to define which side of the plane is on the outside of the object. These points can be obtained conveniently by having the robot touch off the points defining the planes on the actual objects. The difference between the first and second, and the first and third points are taken to produce two vectors which are orientated parallel to the surface of the plane:

$$POP1 = P1 - P0 \tag{16}$$

$$POP2 = P2 - P0 \tag{17}$$

where $P0$, $P1$, and $P2$ are the three points on the plane, and $POP1$ and $POP2$ are the two vectors parallel to the plane. If the cross product of the two vectors parallel to the plane is taken, the resultant is a vector normal to the plane:

$$N = POP1 \times POP2 \tag{18}$$

where N is the vector normal to the plane. This normal vector is the basis for the coefficients A, B and C which form the normalised normal vector for the plane. The fourth coefficient, D, is found by taking the dot product between one of the points on the plane and the normal vector:

$$D = N \cdot PO \tag{19}$$

or,

$$D = Nx\ POx + Ny\ POy + Nz\ POz \tag{20}$$

which is the same form as the plane equation itself, with $Nx = A$, $Ny = B$ and $Nz = C$. Thus, N gives the values for the plane coefficients A, B and C. All of the coefficients are then scaled so that the normal vector becomes a unit vector. This is accomplished by dividing all the coefficients by the factor:

$$SQRT\ (A*A + B*B + C*C) \tag{21}$$

which is the magnitude of the normal. This has the effect of scaling the equation so that a simple addition of a constant to the parameter D will shift the plane by an equal amount in the direction of the normal. This allows a 'safety margin' to be added directly onto D to shift the plane of the volume out of the actual surface. The final step in determining the coefficients is to take the orientation of the half space into account; this is determined by the fourth 'outside point'. The outside point is substituted into the half-plane inequality (Eqn. 15) to see if the half-plane is orientated correctly. If the inequality is satisfied by the outside point then all of the coefficients must be negated to make the inequality agree with the known facts. The effect of negating the coefficients is to reverse the half-space so that what was inside is now outside, and vice versa.

As mentioned earlier, the safety margin is defined as a function of velocity. This velocity dependence is accounted for by modifying the value of D. At present this is done by taking the component of the velocity normal to the plane (obtained by taking the dot product of the velocity and the plane normal), scaling it, and adding the magnitude of the result to D. This has the effect of adding a velocity dependent safety margin to all of the forbidden volumes, effectively causing the volumes to elongate in the direction of motion. The scaling factor is chosen empirically to provide enough stopping distance between the detection point and the actual volume to prevent a collision.

Concluding remarks

The safety computer was initially tested in November 1983. It is now used on a daily basis while other systems are being tested. Although the safety computer has undergone only a limited number of hours of operation, it has proved to be quite reliable and extremely useful during the development stages of other systems. This was particularly true for integration testing of the NBS RCS with other sensor systems, such as the NBS vision system, where movement of the robot was necessary to verify proper operation.

One of the primary requirements that influenced the design of the safety computer was that it must be easy to operate. This is particularly important since personnel who are not intimately familiar with the system will be using it while conducting tests on the T^3. The procedures to start up the safety computer and to recover from an error condition are straightforward. Self-explanatory prompts are displayed on the user terminal making it simple to modify the velocity, acceleration and stopping threshold parameters and to understand what error condition was detected.

A number of additions to the capabilities of the safety computer are already planned. A parallel communications link between the safety computer and the NBS RCS is in the process of being added. This link will enable the safety computer to verify that the RCS is operating properly, and vice versa. The RCS will also be able to send the safety computer the goal point, i.e. the point to which the robot is being sent. Thus, the safety computer will be able to determine that the robot is about to enter a forbidden volume before it actually gets there. This may eliminate the need for a safety margin, or at least permit it to be reduced.

There are plans to increase the sophistication of the forbidden routines by including checks on other parts of the arm besides the tool point, such as the wrist. The current model of the robot used in these routines is a stick figure with pivot points corresponding to each joint. It may be possible to use a solid model of the robot if the computational time is not too great. By using a solid model, a protective envelope could be constructed around the robot arm to prevent sensors and other hardware mounted there from entering forbidden volumes and potentially being damaged. These and other enhancements will be tested in future versions of the safety computer.

References

1. Simpson, J. A., Hocken, R. J. and Albus J. S. 1982. The Automated Manufacturing Research Facility of the National Bureau of Standards. *Manufacturing Systems*, 1(1): 17–32.
2. Barbera, A. J., Fitzgerald, M. L., Albus, J. S. and Haynes, L. S. 1984. RCS: The NBS real-time control system. In *Robots 8 Conf. Proc.,* Vol. 2, pp. 19.1–19.33. Society of Manufacturing Engineers, Dearborn, MI, USA.
3. *Operating Manual for the Cincinnati Milacron T³ Industrial Robot*, Publication No. 1-IR-79149, 1980. Cincinnati Milacron Company, Cincinnati, OH, USA.
4. *Service Manual for the Cincinnati Milacron T³ Industrial Robot*, Publication No. 3-IR-80147, 1981. Cincinnati Milacron Company, Cincinnati, OH, USA.
5. *CPU-Specific Documentation for polyFORTH on the Intel 8086 / 12,* User's Supplement to the polyFORTH Reference Manual, 1980. FORTH Inc., Hermosa Beach, CA, USA.

Increased Hardware Safety Margin through Software Checking

B. A. Cook
IBM Corporation, USA

An implementation of built-in safety features of a hydraulically powered robot, along with additional recommendations for the user to implement at the worksite are described. The main theme is the use of the software control system for most of the built-in safety features, along with an explanation of each. These features include the real-time 20 ms loop axis movement tolerance checking in various configurations, freeze motion mode, startup procedures and emergency hydraulics off. The necessary hardware and its interface to the software will be included to show how each function is implemented.

A robot is an electromechanical machine whose arm can move rapidly with a lot of force under programmed control. There are predictable hazards associated with the operation of a robot, as with any machine. Extreme caution should always be exercised when working around a machine, and particularly when the workspace of the moving parts is entered. In general, no one should enter the work area of a manipulator while power is on. For those limited situations when an operator needs to enter the work area, such as to correct a jam or load a feeder, additional precautions are necessary. For example:

- The operator should never place the head or upper body inside the work area.
- A portable emergency power-off device should be within immediate reach.

There are also unpredictable hazards associated with robot operation. As with any such machine, a system failure can occur, thereby resulting in a potential safety hazard. The system designers can assist the user in providing a degree of protection against these.

The IBM 7565 Manufacturing System has, in addition to standard machine safety hardware, unique safety features built into the software control system. Together, all of these provide increased protection against personnel injury and equipment damage due to hardware failure.

System description

The IBM 7565 Manufacturing System (Fig. 1) is a highly intelligent robotic system designed to perform a wide range of light assembly, fabrication, testing, and materials handling tasks. A typical system consists of:

- A manipulator plus optional gripper. The manipulator is a rectangular box-frame device that supports an arm with wrist and two-fingered gripper. An operator panel is located on the front of the manipulator for controlling startup, motion, and emergency off (Fig. 2).
- System controller. This is an IBM Series/1 computer, modified to include servo circuitry to control the manipulator. A typical controller configuration includes dual diskette drives, a video display terminal and a printer.

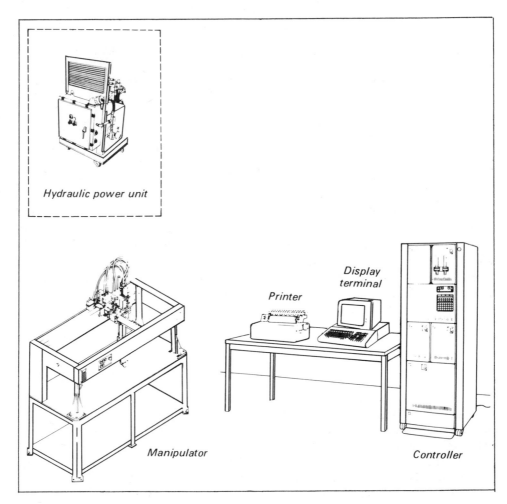

Hydraulic power unit

Printer

Display terminal

Manipulator

Controller

Fig. 1   IBM 7565 Manufacturing System

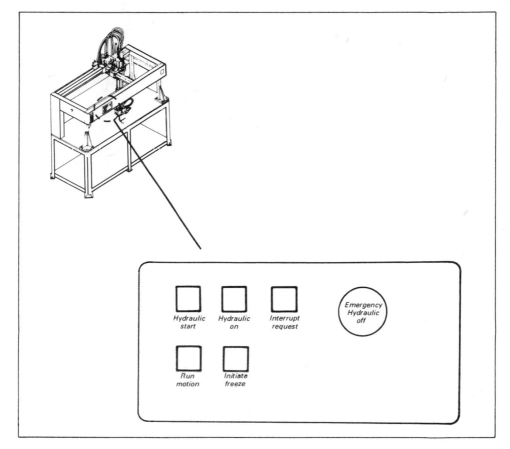

Fig. 2 Operator panel for controlling startup, motion and emergency off

- Programmable teach pendant.
- Hydraulic power supply.
- A Manufacturing Language (AML). This full programming language was designed for manufacturing use. It allows simple, high-level subroutines to control motion, monitor sensors, control peripheral equipment, and perform complex calculations, as well as storage, printer, display and communications management.
- Data communications and remote job entry capability.

Safety hardware

The safety hardware provided with the system includes (Fig. 3):

- An emergency power off switch on the controller/safety timer circuit which shuts off the hydraulic power supply if the controller power is shut off.
- Emergency hydraulic off switches which shut down the hydraulic power supply.

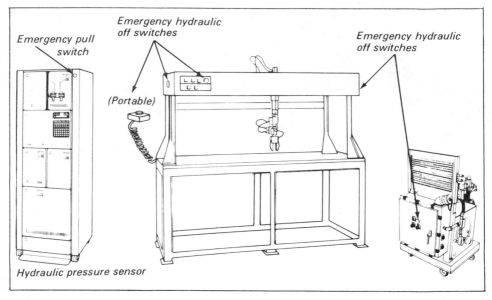

Fig. 3 Safety hardware

- An hydraulic pressure sensor which shuts down the hydraulic power supply if pressure is too low; and issues an error to the display terminal.
- An hydraulic fluid level sensor which prevents hydraulic power supply start-up or causes shut down if fluid level is too low; also issues an error to the display terminal.
- An hydraulic fluid temperature sensor which shuts down the hydraulic power supply if temperature is too high; issues an error to the display terminal.
- An hydraulic power supply thermal overload protector which shuts down the hydraulic power supply if the motor exceeds the allowable temperature limit.
- A service light which flashes when servicing bypasses safety checks.
- Customer reset safety which disables hydraulic power using digital input; this can be used by the customer to connect intrusion devices or safety mats around the manipulator periphery to disable the hydraulics when the safety zone is penetrated.
- An air pressure sensor which issues an error to the display terminal if there is insufficient air pressure to counter-balance the z axis.

Safety software

Manipulator modes

There are three manipulator modes[1]: IDLE, RUN and FREEZE. Manipulator mode determines the allowable motion and defines system integrity checks. Software subroutines provide control over the mode the manipulator is in.

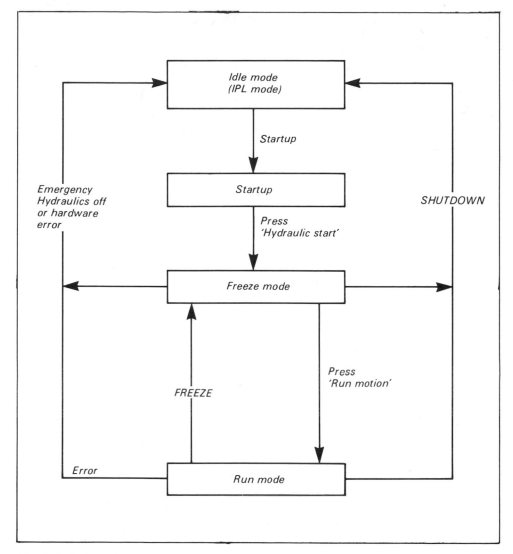

Fig. 4 Relationship between manipulator models

Fig. 4 shows the relationship between the system modes. It can be seen that there are only certain ways that each mode can be entered:

- By system subroutines like STARTUP, SHUTDOWN, AND FREEZE.
- By the operator pressing buttons.
- As result of a system hardware error.

IDLE mode. This is entered at the time the computer controller is turned on, with the execution of the SHUTDOWN system subroutine, or with the occurrence of a system error that causes hydraulics to be shut down. In IDLE mode, the system does not perform any integrity checks nor does it issue any

I/Os unless the operator commands it to. The hydraulics cannot be turned on but motion system subroutines are permitted and execute normally, with the calculated goals being sent to the servo cards. The position of any joint configured in the system can be read; thus allowing the joints to be moved manually through a path and the approximate position of the joints at key locations recorded.

Simulating motion steps in this manner can be useful for debugging programs while providing an inherently safe condition, since hydraulics are off and the joints cannot move. This mode is also useful for executing other data processing and diskette generations. The errors that put the system into this mode are:

- Joint position error tolerance check.
- Manipulator and sensor I/O errors.
- Invalid configuration data.
- Failure to service the operator control panel FREEZE programmable function key.

To get the system out of IDLE mode and into FREEZE mode, the operator must issue the STARTUP system subroutine. Then, to get the system into the RUN mode, the operator must press the RUN MOTION button.

FREEZE mode. This can be entered from IDLE mode by calling the STARTUP system subroutine and pressing START HYDRAULICS, or it can be entered from RUN mode by calling the FREEZE system subroutine. These system subroutines can be called from the display station or from a program. In FREEZE mode, the hydraulics are on but motion system subroutines are not allowed. Because hydraulic fluid flow is reduced by solenoid-operated flow control bypasses in both the pressure and return lines while the manipulator is in FREEZE mode, this is the safest mode that the system can be in while hydraulic power is on. However, the FREEZE system feature is not intended for use in emergency situations. If the FREEZE switch on the manipulator operator panel is pressed, it may take up to 5 seconds before FREEZE mode is entered. The emergency hydraulics off function stops the manipulator arm's motion more quickly than calling the FREEZE function.

While hydraulics are on, the system performs the following integrity checks:

- Joint position error tolerance check which verifies that the joint is within a system-defined distance of its intended position.
- I/O device integrity which verifies that each I/O issued by the system works correctly.
- Hydraulic pressure, temperature level which verifies the hydraulic oil parameters.

If one of these integrity checks fails, the hydraulic pump is shut down, a message is sent to the display station, and IDLE mode is entered.

RUN mode. This is entered only from the FREEZE mode and only by pressing the RUN button on the operator console. In this mode, the hydraulic pump is on and all motion system subroutines are allowed. The same system integrity checks made for FREEZE mode are made for RUN mode. If one of these integrity checks fails, the hydraulic pump is shut down, a message is sent to the display station, and IDLE mode is entered.

Tolerance checking

Associated with joint motion is a check that monitors the progress of a joint's motion in relation to the calculated motion[2]. This check is referred to as tolerance checking and is active when hydraulics are on. Tolerance checking provides both safety and integrity.

The system detects tolerance errors for both moving and stationary joints. When joints are stationary, the system allows them to differ from the commanded position by an amount intended to cover only small deviations that could be expected under normal conditions. In other words, when a joint is stationary, tolerance is very tight. When a joint is moving, the system allows it to differ by the maximum tracking error for that joint. The gripper tolerance is large to accommodate the large error signal required to exert pressure on a held object.

Startup integrity checks

The AML subroutine STARTUP causes the following sequence of events to occur to verify the integrity of the position sensors and other control system electronics before allowing the hydraulic pump to be started by the operator:

- Finds the arm position.
- Issues a position offset to ensure feedback tracks the command.
- Enables the pump start relays; monitors for pump enable feedback.

Once the pump is started, additional system integrity checks are made by the STARTUP subroutine:

- Monitors customer reset safety.
- Monitors position to ensure no excessive drift occurs.
- Monitors hydraulic pressure to assure an acceptable level is obtained within a prescribed period of time.
- Issues a position offset and monitors the error signal to assure the arm tracks the command.

Failure to obtain corect results causes the pump to be shut down. The subroutine allows the system to go from IDLE to FREEZE mode if all system checks are met.

References

1. *IBM 7565 Manufacturing System Software Library, A Manufacturing Language Concepts and User's Guide,* 1983. International Business Machines Corp., Boca Raton, FL, USA.
2. *IBM 7565 Manufacturing System Hardware Library, General Information and Site Preparation Manual,* 1983. International Business Machines Corp., Boca Raton, FL, USA.

Training and Design for Safe Implementation of Industrial Robots

L. R. Carrico
International Flexible Automation Center (INFAC), USA

To provide the maximum knowledge in reference to a safe and effective implementation of industrial robots, the engineering personnel involved must be properly familiarised and trained on the robotic equipment to be used. Continual awareness of safety and the 'man-machine' interface is a must. An overview of the criteria to follow when performing the task of designing and implementing robot applications is presented.

The utilisation of industrial robots in industry has brought about the need to take a closer look at the safety criteria regarding man and machine. Because of the many repetitious and accurately controlled movements performed by industrial robots, personnel involved with the equipment tend to become overly familiar and careless. Personnel are still not totally aware that industrial robots are *not* just another machine that they have normally worked with – they are 'reprogrammable multifunctional manipulators' by definition, not standard, stationary equipment with guards, doors and totally visible paths of motion. In addition, it must also be realised that we are still in the era of the 'dumb robots'; that is, the robots of today do not have standard capabilities of perception and senses built-in.

Future generations of robots will no doubt have these 'inbred' capabilities. However until designs achieve these levels, engineering personnel need to provide these 'senses' through the use of the software features, interlocks, etc. made available by the robot manufacturers or incorporated on-site by the user's engineers.

The problem still facing us today is the small percentage of personnel who are knowledgeable in robotics. This lack of knowledge has caused the implementation of robots to be lengthy, problematical and costly. To provide the maximum knowledge in reference to a robot's features and capabilities, the first and most important consideration should be the familiarisation and training of all personnel. Once properly and formally trained, a user's engineering staff, along with the vendor's assistance and/or recommendations, should have the capability to design an effective and safe robotic application.

After effective training has been accomplished, the next step is obviously the implementation of the design – this is probably where most safety factors are overlooked. When installing the equipment, there invariably will be changes and / or modifications to the design. In some cases, these changes are not totally researched or documented and, as a result, may alter the safety of the application area as originally designed. Thus steps need to be taken to add additional safeguards.

Another problem during implementation is the assumption that personnel will readily stay clear of the area, thus the user's engineers must continuously be aware of maintaining safe clearances when designing the application.

Training and familiarisation with equipment

The most effective way to utilise any equipment is to be totally familiar with its capabilities and features. Thus, the need for effective training!

In many past cases, the personnel required to design a robotics application had little or no knowledge of the equipment or the job to be performed. As a result, problems occurred when attempting to provide a safe work area and safety-orientated interfaces, and, in some instances, 'pinch-points' were created (those locations where personnel could be trapped between the robotic arm and peripheral equipment or fixtures).

To prevent these and other situations recurring, the user's engineers responsible *must* be properly trained. The first area to consider is familiarisation with the type of robot to be used. If an engineer is familiar with one type of robot, it does not mean that he is knowledgeable with all types of robots. He *must* be made aware of the type of robot design he is to work with (rectangular, cylindrical, spherical or anthropomorphic), its type of control system (point-to-point, continuous path or controlled path), its axes of movement, ranges of motion, path velocities, and so on.

Next, he must be instructed in the capabilities of the control and its features. One of these features is the type of pendant or teach mechanism used, such as a push-button or a microprocessor type. The user's engineers should also become familiar with the input/output interface structure. This will be his primary tool in designing an effective and safe application.

It must be remembered that robots available today have emergency stops to provide total shutdown, and most have interrupt type features to cease robot movement while maintaining power when an application area is entered. What most robots do not have are hardware devices standardly available to sense when personnel or equipment is in the robotic work area. These features need to be designed-in by the engineer in reference to the application.

Control systems for robots usually work with a priority interrupt scheme. Interrupts of higher priorities are those reserved for hardware devices to signal that personnel are entering the work area or that tooling needs to be changed or repaired. In these cases, the hardware device can be utilised to direct the robot to a safe area through a pretaught sequence. This safe area would allow maintenance personnel to work on attached tooling from the outer edge of the robot's reach. Thus, in case of a control or robot malfunction, the maintenance person would be beyond the work volume (range-of-motion) of the robot.

In addition, most state-of-the-art controls have error codes. These error codes have been reserved as the highest interrupt priorities for the monitoring of the software system controlling robotic arm movement. When ambiguities in the feedback signals of the servo systems are sensed, these interrupts will provide a total robot shutdown, preventing a machine 'runaway'. With the possibilities of heavy loads continuing to move because of inertia, hardware stops are also incorporated in the robot design as a backup. These hardware stops are normally set for maximum range-of-motion capability of the robot when the robot is manufactured. Engineering personnel designing an application must ensure that these hardware stops are moved to reduce the work volume of the robot if total range-of-motion is not required for the application.

The user's engineers should also be familiarised with the control protection features in reference to the executive software and data program. What type of memory is being used? Is it volatile? If so, is there a back-up and for how long?

State-of-the-art processors are being utilised by most robot manufacturers today. Some are designed with ROM (read only memory) and some with RAM (random access memory). RAMs are volatile – the software program would be lost when power is turned off or fails. To prevent this, back-up battery units are built into the controls. Upon turn-off or loss of power, the back-up battery provides the current needed to keep memory refreshed. Some manufacturers have added insurance by having the current to be supplied come directly from the prime power source; this provides the capability of turning off the control at the end of the production day without loss of memory, and without using the back-up battery. The battery is kept at full charge through another current source from prime power, thus keeping it in reserve for prime power losses.

Once familiar with the robot's design, range-of-motion, the control system, input/output structure, and software protection features, the user's engineers should then be ready to design the application.

Application (system) design

When designing the robot application, engineering personnel must remember to utilise the features on which they have been instructed, and should follow some basic guidelines, such as:

- Start with a simple application (if this is the first).
- Communicate with shopfloor personnel.
- Allow time for proper training and project debugging.
- Use vendor support.
- Remember 'human' engineering.

Equipment layout
It should be remembered to look at the entire job, work area to be used, and the peripheral equipment involved. A prime consideration is to prevent any 'pinch-

points.' This can be accomplished by using templates and drawings of the equipment to place them in the optimum position for completion of the application. It must be ensured that the work area is designed so that there are no obstacles in the path of the robot – remember, a standard feature of most robots is not sight. They have to be able to perform a designated task without obstructions in their paths of motion.

In addition, proper clearances must be allowed when the robot is working with peripherals, be it machine tools, presses, transfer lines, palletisers, gauging stations, and so on. Even the best conceptual design can end up as a lot of wasted effort by failure to do so.

Insurance must also be provided to allow proper clearance and safe areas for the most important factor of all – personnel. Although robots perform repetitive tasks for long periods of time, they are still machines and will require periodic preventative maintenance and may eventually require repairs due to devices, modules and/or tooling working incorrectly. As previously mentioned, clear spaces and safe areas *must* be incorporated in the design. There may be some jobs that will require work to be done with power on (such as alignment and repair of servo systems). This must be realised in the design

Fig. 1   Example of front-loading of machine tools

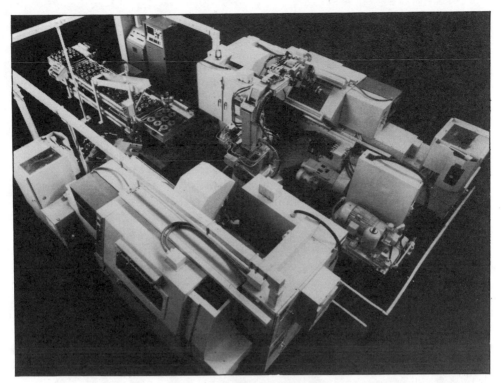

Fig. 2   Rear-loading/unloading of machine tools allows maintenance and engineering personnel to work on equipment with greatly reduced danger concern

stage by the user's engineers. It cannot be assumed that work will always be performed in a power-off state.

For the dressing of tooling tips, and tool repair or replacement, space must be allocated so that the personnel involved can perform their job beyond the robot's range-of-motion.

When planning the placement of equipment, a key factor to keep in mind is that of how to place the robot in reference to a peripheral machine. It may be beneficial to have the robot load, or work with a machine from the rear to have a safer application. An example is machine tool loading applications. Initially, designs were implemented that had the robot load a machine tool from the front (Fig. 1). This made very good sense, as the door to the machine was located there. To provide maximum use of the robot, and to increase production, another machine tool was added to the application, creating a manufacturing cell. But one major problem presented itself—that of apprehension on the part of personnel performing maintenance on the cell.

Features were used to lock-out one machine tool to have maintenance performed, while the robot loaded/unloaded the second machine tool. Even though personnel knew that the lock-out was incorporated and worked, they were very concerned for their safety with a robot moving behind them! The solution was to have the machine tools turned around, doors designed in the back, and thus loaded/unloaded from the rear (Fig. 2). The end result was a much safer cell design, and it allowed maintenance and engineering personnel to work on equipment with greatly reduced concern of being in danger from the moving robot. This concept can and should be used in considering similar types of applications.

Perimeter guarding

Having designed the equipment layout for the application, the next concern of the design engineer should be to provide for the isolation of the work area by using some form of perimeter guarding. Effectively designed, this will prevent non-involved personnel from using the work area as a 'walkway' or short-cut to other areas.

It may be desired to design this perimeter guarding so that if the work area is breached, the robot and peripheral equipment will go into a 'hold' or 'interrupt' status – the cycle of operation would be stopped to prevent the hazard of persons being in danger from a moving robotic arm. Power would still be applied, thus allowing for continuation of the cycle of the operation from that point in the program after the area is clear, in lieu of reinitiating the operation from the cycle start point.

Fig. 3   Perimeter guarding using fencing with gates

A wide variety of perimeter guarding and devices are available, including chains and guardrails, fencing with gates, light curtains, and pressure sensitive mats.

Chains and guardrails are very useful in surrounding the work area. They readily eliminate equipment such as forklifts and hand-trucks from being moved through the work area, and deter some people from walking through the work area. But it is recommended that they be used with reservation. People are 'creatures of habit'; if the work area was used as a short-cut walkway in the past, personnel may tend to climb over the guarding and continue to travel through the area. Therefore, it is recommended that guardrails be used along with one of the following other types of devices to ensure proper and total isolation of the application area. (See also page 181).

Fencing with gates is a very effective and fairly inexpensive design for perimeter guarding (Fig. 3). The gates should have some type of device or switch incorporated (such as a proximity switch) to generate an interrupt signal to the controls of the equipment. When the gate is opened, the robot and peripherals would then go into the 'hold' or 'interrupt' state. A second device should also be considered in the design along with the first device, such that they both must be maintained to reactivate the equipment. This will eliminate personnel from going into the work area, closing the gate, and inadvertently reinitiating the cycle of operation. It is advisable to locate this second device outside the fencing so that on leaving the work area, personnel must go to it for activation, thus preventing them from being present in the work area when the cycle of operation is restarted. The fencing should also be designed high enough to prevent personnel from climbing over it. A height of approximately 2.4 metres should be considered – personnel would not be able to scale it without some effort; if they do attempt to scale it, it should hopefully take them long enough so that other personnel would see and stop them. (See also page 217).

In some applications, light curtains are used to safeguard the area (see also page 199). As long as the light curtain is maintained, the cycle of operation will continue. When personnel or equipment 'break' the curtain, the signal is generated for an in-cycle 'hold.' These light curtains should also be used with discretion. Although effective, the environment of the work area may cause the intermittent generation of the 'hold' signal caused by foreign 'particles' being generated in the area from the application process. The work area should be a relatively 'clean' environment, to prevent this intermittent 'hold' signal from being generated.

The fourth example to be discussed is pressure sensitive mats (see also page 205). These should be installed in the area encompassing the robot and peripheral equipment; but, it should be ensured that they do not extend too far beyond the outer perimeter of the work area. This could cause the generation of the hold/interrupt signal at unwanted times.

Pressure sensitive mats cause a signal to be generated when weight is applied. The user's design engineers should ensure that the work environment is relatively clean and clear of non-related items. These items inadvertently placed on the mats would also cause unwanted stoppage of the operation cycle.

As mentioned previously in reference to guardrails, this type of safeguarding should not be used by itself. A second device should also be incorporated some distance from the work area such that it must be made along with the absence of weight on the mat to reinitiate the cycle of operation.

Interfacing to remote stations

When designing a safe robotic work area, an important piece of equipment to consider is a remote station or operator panel. This panel would have remote emergency stops and hold/interrupt features in addition to those made available on the manufacturers equipment, and in addition to the switch devices designed in the guarding fixtures. The panel should also have the second device needed to reinitiate the cycle of operation, as previously discussed. Placement of the panel should be such that it is not in the immediate work area but in close proximity to allow for quick corrective action to take place (Fig. 4).

In reference to emergency stops, it is also advisable to have these provided at several locations around the work area. Their location should be designed with no obstructions in the way, so that easy access is attainable from any direction.

Floor marking

Having designed the layout of the equipment in the application area, the perimeter guarding, and the interfacing to remote panels, it is necessary to

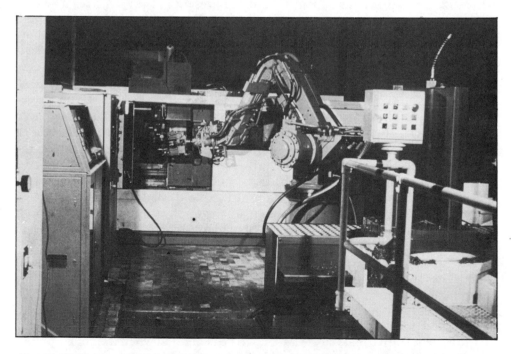

Fig. 4   Placement of the operator panel (or remote station) should be outside the immediate work area but close enough to allow quick corrective action

delineate the robot's range-of-motion with floor markings. This will provide an immediate visual reminder to the personnel in the area. These floor markings should be made of long-lasting paint or tape, and should be of a colour that meets OSHA safety standards. The most commonly used colour is yellow, signifying 'caution'.

Implementation

To ensure that the installation takes place as planned, it is imperative that the engineer has a complete documentation package available. This documentation, including layout drawings and electrical and mechanical interface prints, would be developed during the design phase. The documentation should also contain a description of the cycle of operation to include peripheral equipment, operator panels and special interfacing.

During the installation, there will probably be design changes made. Before implementing these changes, they should be researched in reference to their impact on the safety criteria of the original design. Once safety has been maintained, these changes should be fully documented so as to have an up-to-date record of the application area.

One of the most important factors to remember during implementation is to allow enough time for the proper instruction of the personnel involved on overall system operation. This would include engineers, maintenance personnel and any operators. It should also include representatives from the facility's safety committee or group to ensure that all regulations have been adhered to. This can be accomplished by the vendor's representatives and the technical training coordinators of the user.

Remember the key factor of a successful and safe application is the proper indoctrination and training of *all* personnel involved. This will make them aware of the important safety devices/features that have been incorporated in the application work area.

It is also important to ensure that refresher seminars be conducted. Because of the high reliability of today's equipment, people may tend to forget or ignore some of the key factors of the application. These refresher seminars should emphasise the cycle of operation and the safety characteristics, and should include applications sessions to exercise skills learned, thus providing them with an up-to-date awareness.

4
System Components

This section deals with the individual items which may be used as part of the robot safety system. These include perimeter guarding, safety interlocking, photoelectric guarding, safety mats, safety sensor systems and a robot safety and collision avoidance controller.

The first paper by Robinson provides an introductory overview of the problems of guarding. He considers each of the three National Bureau of Standards 'zones of safeguarding vigilance': level 1 the workstation perimeter, level 2 within the workstation and level 3 adjacent to the robot arm, each having its own specific requirements. The author concentrates on levels 2 and 3 – the 'neglected zones' – whereas level 1 is covered in the later papers of this section.

Specifically, Colin Thompson previously of the Ford Motor Company discusses safety interlock systems noting that the interlock system may be hydraulic, pneumatic, mechanical, electrical or any combination of these and must not 'fail to danger'. Limit switches, plug and socket systems and captive key systems are described in detail.

The next paper by the Manufacturing Safety Coordination of Ford of Europe provides information based on the experience in plants where photoelectric guarding has been used for some time. Following this safety mats are discussed in detail by Mike Graham of Herga Electric who after dealing with the safety mat design and construction then examines applications in robot installations. The fifth paper of this section by Brian Powis of J.P. Udal Co. discusses the construction and requirements of perimeter guarding of robot installations.

The last two papers are somewhat different. First, Kilmers' discussion of safety sensor systems is leading into a relatively new but extremely important area. The sensor systems are broken down into the three levels, Level I – perimeter penetration, Level II – intruder detection within the workstation, and Level III – intruder detection near the robot. Likewise he discusses three general types of security intrusion detection system: point systems to detect an intruder at a single location, perimeter systems to detect across a boundary, and area or space systems to detect intruders within a selected region. A prototype safety system designed for the Stanford Arm robot is also described. It is anticipated that the application of sensors will become progressively more important over the years ahead.

The final paper of this section by Derby et al. of the Rensselaer Polytechnic Institute is an examination of the use of four different sensor types (ultrasound, microwave, infrared and capacitance), and how they might be linked to an integrated system under the control of a central microcomputer. The paper ends with a brief discussion of what needs to be done to introduce the system onto the shop floor.

Robot Guarding – The Neglected Zones

O. F. Robinson

The US National Bureau of Standards identifies three 'zones of safeguarding vigilance' around industrial robots: the perimeter, the zone within the perimeter, and the contact zone immediately adjacent to the robot arm. Perimeter guarding receives a great deal of attention, but the majority of accidents occur to fully authorised personnel within the perimeter. Combinations of area presence detectors and collision sensors are required in addition to perimeter fencing.

It is a fact beyond dispute that the most certain method of saving human beings from injury by robots is to ensure that personnel cannot be present within the operating zone. The methods of doing this – perimeter fencing, interlocked guards, trapped key gate locks, all combined with safe systems of work – are covered elsewhere in this section.

This paper concentrates on an aspect of guarding which has been less well aired, that of protection of personnel – often fully authorised – working inside the perimeter fence, beyond the access gate.

Experience in Japan, North America and Europe all suggests that the vast majority of accidents involving robots occur not during normal operation, but during teaching, programming and maintenance. During these operations the robot should be functioning in teach mode, but there is still the possibility of hazard caused by aberrant behaviour of the robot as the result of control or mechanical failure, or caused simply by operator error or negligence.

In addition to hazards for those people who are permitted access to the operational zone, there is the problem of personnel straying inside the perimeter through positions where the integrity of the fencing is breached by the need for access of transfer lines or overhead conveyors. To cite an example illustrating this possibility, last year (1984) in the USA, a production line worker, motivated by the wish to increase his production credits as the end of his shift approached, left his workstation and moved up the line. No fence or light barrier could be used to prevent this as such a barrier would have inhibited the movement of the workpieces on their conveyor. The worker unwittingly stepped into a pinch point and was trapped and crushed by the transfer line.

Even less comprehensible is the recent case, in the UK, where a worker went to considerable trouble to scale what appeared to be a perfectly adequate perimeter fence. Nobody knows why he did this – he was killed before he could explain.

The zones of safeguarding vigilance

The US National Bureau of Standards define the hazard areas associated with robots in terms of three 'zones of safeguarding vigilance' (see also page 223):

- Level 1 is the workstation perimeter.
- Level 2 is within the workstation.
- Level 3 is adjacent to the robot arm.

Each of these zones has specific requirements in terms of guarding.

Level 1, the perimeter, demands a form of guarding which comprises either a physical or a trip-switch type of barrier, a combination of fences, gates and, perhaps, light curtains. The barrier prevents movement beyond a specified point, or detects movement passing that point. Level 1 guarding must, for programming and maintenance reasons, permit the operation of the robot under controlled conditions, with authorised personnel positioned beyond the barrier. It may also, as we have seen from the examples already quoted, be almost impossible to guarantee that the barrier cannot be defeated in an unauthorised manner, by someone sufficiently determined to do so.

Level 2 safeguarding poses an even more difficult problem – the sensing of the presence of humans anywhere within the huge volume of free space often 4000 cubic metres or more – in which the robot moves.

Level 3 demands the capability to detect the presence of a person or other obstruction in direct contact with, or in immediate proximity to the robot arm, and to arrest the movement of the robot until the hazardous situation has been signalled as clear.

It is Levels 2 and 3 – the neglected zones – on which this paper concentrates.

Level 2 – Within the perimeter fence

When planning the method of guarding for Level 2, the most significant feature compared to perimeter guarding is that the emphasis is on detection rather than on a physical barrier. Of course, under some circumstances, presence detection usually in the form of a light curtain, may be used instead of a fence at the Level 1, perimeter stage. In this case, the sensing or trip-switch device is operating in the vertical plane – it is forming an invisible wall. In the case of Level 2 guarding, the sensing device has to operate in the horizontal plane – it is covering a ground area, not a single passing point.

One method of meeting this requirement is to simply take the light curtain and direct it horizontally instead of vertically. The space taken up by the opto-electronic emitter and receiver panels may pose problems in some cases, but in general such a solution seems to work reasonably well. Horizontal light guard systems set around waist level will be less prone to accidental damage. Because it has to detect narrow ankles rather than more bulky waistlines, a low set horizontal system will require closer-set beams and will therefore tend to be more expensive than a waist-level device. Opto-electronic systems, their installation patterns and maintenance procedures should conform to Health and Safety Executive Guidance Note PM41 and the British Standards Institute Document BS6491.

The alternative to light curtains is to use a pressure-sensitive mat, usually referred to as a 'safety switch mat.' (See also page 205). Traditionally, these mats have been very similar to those used for the familiar purpose of opening and closing automatic doors. The main difference has been that the versions used for machine guarding purposes have a degree of self-monitoring within their control system to ensure failure to safety in case of malfunction. Those pressure sensitive mats which are regarded by the safety authorities in the UK as having the necessary failsafe control systems, operate either by means of a matrix of low-voltage electrical switches, or by a web of pneumatic tubes and valves. These are sandwiched into a mat or carpet which is connected, via an interface, to the machine control circuit. Pressure on the mat causes a valve or relay to operate, thus providing a signal to the machine control system.

A major disadvantage of switch mats has been their susceptibility to damage. They are very vulnerable, positioned to the floor of a manufacturing area. Their flexible rubber, PVC or composite surfaces may be penetrated by the sharp corners of handtools of workpieces dropped onto them. They are also susceptible to damage by ingress of dirt, water, oil, solvents, etc.

But one of the major problems of Level 2 guarding – a problem which until recently few users had solved – is that the space within the workstation very rarely comprises a neat, rectangular area corresponding to the shape of an opto-electronic barrier array or of a switch mat.

The volume and shape of space within the workstation area can vary enormously according to the type of application involved. Some means had to be found to provide presence detection within a relatively large area, often of irregular shape, using a detection device which is resistant to environmental hazards. The environmental hazards will usually include impacts from dropped tools or materials and may also include one or more of: liquid spillage of various kinds, welding sparks of hot swarf, exposure to paint or solvent mist, and occasionally, the incursion of materials handling vehicles of various kinds.

These needs have led to the recent development of switch mats made up of relatively small modular sections which can then be joined together and arrayed over large areas of floor. The modularity makes these systems easy to transport and to install. They also have the advantage that if they are damaged, the affected section can usually be quickly and economically replaced without changing the complete system. Modular mat systems are also available with steel surface plates to prevent penetration of the unit by dropped tools or other sharp objects.

Care must be taken in choosing the best system for a user's particular environment. Some units have very efficient sealing against liquid spillage;

others are not sealed at all. Some models are sensitive right to the edge of the individual section, providing presence-sensing at any point on the layout. Others have dead areas around the edges. The degree of importance attached to all of these features varies according to the nature, size and shape of the system and of the process it is guarding.

These modular units, though new to the market, still operate on the same principle as the older integral rectangular mats; that is, pressure on the surface either activates an electrical switch or is sensed by a pneumatic tube which activates a valve. Some of the latest systems under development desert the traditional methods in favour of slabs of flexible plastic material which indicate depression of their surface by a measurable change of resistance in the electrical charge within the material itself. One such system is already on the European market though, in the UK at least, its distributors state that this technology is not yet available to the same level of failure-to-safety control that is possible with the more traditional methods. Despite its current limitation, this system has the advantages of lightness, robustness and the possibility of being cut into any shape. These features could make this system extremely attractive to users if an acceptable control can be developed to ensure failure-to-safety.

Level 3 – Contact detection
Level 3 is described as that part of the working environment within physical contact distance of the robot arm. In many cases, the methods already discussed – switch mats or horizontally arrayed opto-electronic systems – could also be used at Level 3. However, there is an alternative or perhaps additional method of 'last-ditch defence' which has the advantage of versatility, simplicity and relatively low cost – the use of collision detector strips mounted on the buffer arm itself (Fig. 1). They work on a similar principle to switch mats and comprise either a low-voltage electrical strip-watch, or an air-filled tube. When depressed by contact with an obstruction, an impulse from the detector will cause, via a suitable interface, an emergency stop of the robot system.

Collision detection strips have an advantage over other methods of Level 3 guarding in that, where there are fixed obstructions such as pillars or other machinery within the sweep of the robot arm, the detector device can protect the robot itself, under conditions of aberrant behaviour, from self-destructing on the obstruction. For this purpose, an alternative to fixing the detector strip on the robot arm could be to fix it to the permanent obstruction.

The electrical type of detector strip comes in a wide selection of versions. The basic switch can be made up of normally open electrical contactors which 'make' when the outer sleeve is depressed. Alternatively it can comprise a wire or wires encased in an outer sleeve which operate by flexing. The sleeve may be PVC or rubber. In some applications it may be desirable to have a protective outer coating to guard against weld spatter.

There are two critical aspects of the specification of a collision detection device. First, the circuitry must be so designed that, when used in conjuction with the correct controller and interfaces, the system will fail-to-safety if a fault condition arises. Thus an electrical system should detect and shut down if there

Fig. 1 Buffered safety edge switching mounted on the arm of a welding robot

is a break in supply, a short circuit, an overload, or a component failure. An air system must, in addition, monitor the air pressure.

Secondly, the detector should be sufficiently sensitive to permit the robot arm movement to stop before any damage is caused. This implies that the detector switches must be surmounted by a cushion or buffer which is sufficiently deep to take up any pressure until the arm movement ceases. Just how deep this cushion need be will vary according to the robot involved and the individual circumstances of the application.

A potential buyer of a collision detection system would be well advised either to choose a supplier with several different models in its range or, alternatively, to investigate the products of more than one supplier. Some collision detection strips are sufficiently flexible to be able to operate when mounted on a curved surface, or to be angled around corners. An important feature on the larger robot arms is that the detector switch should operate even if the contact is made by a glancing blow from the side. Where the environment contains liquid sprays or dust, the mechanism must, of course, be resistant to this and, where there is the presence of solvent-based paint spray or other potentially explosive material, a non-electrical system would be preferred.

New developments

In addition to the systems currently to be found in the UK, there are a number of interesting variations in use or under development elsewhere, though not (in their present form) considered fully acceptable by expert opinion.

One technique, pioneered in the USA, sets up an invisible radiowave pattern around the guarded area. The system consists of two basic elements. The controller generates the output signal and electronically monitors the responses. Signals are transferred via the second system element, the antenna/sensor (usually abbreviated to 'antensor'). The antensor may consist of any electrically conductive material of suitable rigidity and surface area. Typically ¾ in. diameter steel thinwall conduit tubing is used as this is readily available and easy to bend into any desired shape. This is fixed to the machine to be guarded by means of stand-off insulators. The connection of the antensor to the controller is by means of coaxial cable.

Whilst recommended by the manufacturers for a wide variety of applications, the particular relevance of this radio frequency system to industrial robots is that the antensor can be mounted on the robot arm – what may be described as an electronic 'ring of confidence'. Furthermore, the range of the radio beam may be adjusted so that the sensing distance from the robot arm can be varied according to circumstances.

This type of system must, because of its flexibility, have considerable initial appeal. However, there are disadvantages which would, despite compliance with the US, OSHA and ANSI regulations, make such a system unlikely to be accepted by a British factory inspector. Firstly, it is a comparative system. That is, the system is initially adjusted to a particular set of environmental conditions and signals a stop when the prevailing conditions exceed the norms acceptable to the program. The conditions being monitored in this case are the volume of grounded mass within the range of the antensor. There is always the chance that sensitivity readjustments must be made for drastic changes in conditions, perhaps the distance of the operator from his ground. This introduces a human element which is always fallible.

Secondly, the system is electronic and monitoring is by a microprocessor. There is still a reluctance to accept solid-state technology for machine guarding in the UK and, indeed, throughout much of Europe. Nevertheless, the principle is sufficiently interesting for developments to be kept under surveillance by the various appropriate authorities and no doubt improved versions of the technology will eventually find themselves fully accepted and in general use.

As mentioned previously single unit switch mats detect presence by compression of the mat material which causes a measurable change in resistance. Clearly, the same technology can be used to form a collision detector fixed to a robot arm or other pinch point. The simplicity of the switch construction, allied to its robustness, makes this an interesting system for the future, though still meeting with resistance today for similar reasons that make the radiowave system unacceptable.

The controller

A critical factor with all detection systems is the controller, or interface between detector switch and machine circuit. The debate is in full swing regarding the pros and cons of software stops against a straightforward old-fashioned cut of supply to the prime mover in an emergency situation, and regarding the benefits, or otherwise, of solid-state as against relay technology for guarding applications generally. This is too wide, complex and, perhaps, sensitive an issue to be discussed here. Suffice to say that the manufacturers and suppliers of contact-detector systems can generally provide whatever type of controller is appropriate and jointly acceptable to the user, his machine supplier and HM Factory Inspectorate.

In some applications, it may be appropriate to use different levels of action according to the zone of the intrusion. One light curtain manufacturer demonstrates that breaching the curtain activates, at one level, a synthesised speech warning, at the second level a software stop and at the third level a hardware stop.

Concluding remarks

Throughout their literature, the Health and Safety Executive in the UK emphasises that whatever general guidelines on machine guarding may be published every application has its individual features and must be judged accordingly. If this is true of traditional machine tools, how much more true it must be of industrial robots and flexible manufacturing systems.

No two situations are identical. In many cases the most suitable guarding, providing security with productivity, personal safety with ease of access, will comprise a combination of different methods of guarding.

Fencing, protected locks, trip switches, light curtains, collision and trap sensors and area intrusion sensors all have their parts to play. All combined with adequate training and retraining and safe systems of work which, in addition to being well conceived, must be effectively enforced; experience shows only too clearly that the person least conscious of safety procedures is often the person those procedures are primarily designed to protect.

The value of using a guarding system which requires no maintenance or adjustment should be emphasised – every service provides the opportunity for a potentially dangerous human error on the part of the maintenance engineer or examiner.

Many suppliers of guarding only offer a very limited range, albeit of excellent systems. However, specifiers of safety systems should be aware of the wide range of products now on the market and consider how these can be combined for maximum effect. Therefore, consulting a multi-product supplier, or several potential suppliers, is recommended.

Finally, when working with robots it should be remembered that the greatest likelihood of an accident occurring is in the two inner zones of safeguarding vigilance, and that the victims are much more likely to be fully authorised maintenance personnel or programmers than unauthorised intruders. In order to avoid damage, both to valuable staff and to valuable robots, it is essential to ensure that these inner zones are equipped with presence sensors and collision detectors.

Safety Interlock Systems

C. Thompson
Ford Motor Company, UK

Access to robot enclosures for programming, maintenance, etc., should only be undertaken by trained authorised persons. Such access should only be gained through interlocked gates. Applications of various types of machine guard interlock systems are discussed together with their construction and operation.

The Factories Act 1961 places on the occupier of the premises an absolute duty to fence securely dangerous parts of machinery, so affording to workers effective protection from danger of contact with the dangerous parts of machinery.

The first point to establish is whether or not the robot and its associated machinery (if any) are capable of causing injury. If the power, size or use of the robot suggests that there is a risk of injury from its operation, then safeguards must be provided.

It is generally recommended that fencing acting as distance guards should not be less than 2m high, unless an equivalent level of safety can be achieved using other safeguards. When determining the type of safeguards to be implemented, consideration must also be given to safe means of access and egress for both the workpiece and the employees who will need to program and maintain the robot.

The load/unload stations should be so designed as to prevent operator access to the dangerous parts of operating machinery. Any load/unload opening through which an operator could pass, should be safeguarded by safety devices such as photoelectric safety devices (see page 199), or pressure sensitive mats (see page 205), interlocked to the robot cycle.

Access to the robot enclosure for programming, maintenance, etc., should only be undertaken by trained authorised persons. Such access should only be gained through interlocked gates. It is important that whatever interlocking system is selected, it should operate on all access gates. Once the gate has been opened, the robot and associated machinery, should not be able to operate in automatic cycle until the gate is closed, and the machine control reset. Closing the gate alone should not initiate a cycle start.

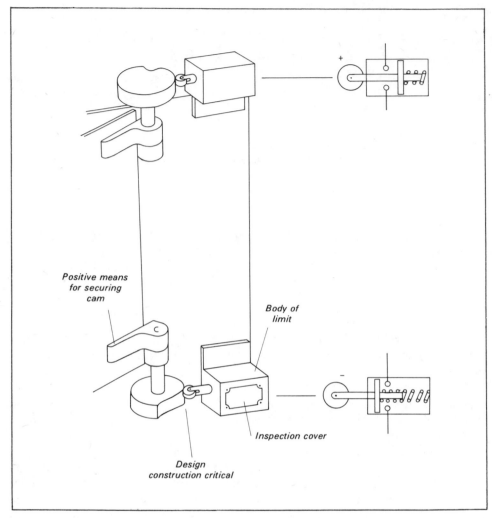

Fig. 1   Typical limit switch arrangement, cam on/cam off

Machines which cannot stop immediately, and therefore have a run down time, must have a time delay mechanism fitted to the gate interlock, which will prevent access to the moving, dangerous parts of the machine.

The interlocking system may be hydraulic, pneumatic, mechanical, electrical or any combination of these, and must not 'fail to danger'.

Limit switches

Limit switches are suitable where frequent access is required as in the case of manual loading/unloading of components on machinery. Where spindle rundown exists, such that contact may be made with dangerous parts on opening the guard or where it is likely that tool breakage/machine damage may be incurred if the guard is opened during the cycle, then a shotbolt arrangement is required in addition to the limit switches.

Limit switches may be positioned to operate on linear or rotating cams controlled by the guard door (Fig. 1).

Direct access is available at each gate. Opening the gate operates the limit switches and isolates the power to the machine parts that may be a source of danger to the operator. Where a shotbolt is required to secure the guard in the closed position until all dangerous movement has ceased, consideration should be given to the fitting of suitable timers/solenoids.

Only safety-type limit switches should be used for machine guard interlocking. Such switches are available from some manufacturers as standard items. When damaged or worn they should be replaced by safety-type limit switches only. Whilst difficulties could arise in the identification of safety-type limit switches, this has been addressed by some suppliers in that they paint their safety limit switches red. However, it does not follow that all red limit switches are of the safety type, and the only positive method of determining a safety limit switch is by a visual check inside.

For high-risk applications safety switches are normally coupled to the guard in pairs (make and break). Only direct action limit switches should be used for interlocking guards. Fig. 2 shows a direct action limit switch operating in the positive mode. The switch contacts are opened positively and closed by spring action, thus forcibly parting the contacts if they become welded together. Fig. 1 shows two limit switches arranged to operate in opposite modes, i.e. one normally open contact and one normally closed. However, if one switch should fail it will not be immediately apparent and to avoid the operator's safety being dependent on the correct functioning of the other, some means of monitoring the switch circuit should be incorporated.

Extreme care needs to be taken when establishing the design and construction of the cam to prevent overrun or incorrect location of the limit switch plunger or arm. The limit switch should be arranged such that it operates before the gate can be opened sufficiently to allow access, as this could

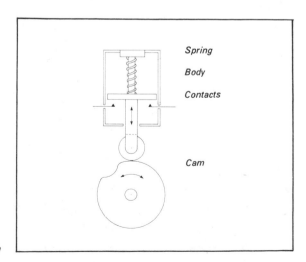

Fig. 2   Direct action limit switch

result in access to live dangerous parts in a guard-open condition.

Careful consideration must be given to the design and construction of guards for the reliable operation of switches. Immediate identification of safety switches as opposed to conventional micro switches can be difficult. Although the initial cost of a dual switch arrangement is favourable in comparison with other systems, this benefit is lost when solenoids/shotbolts are necessary to secure guard doors until rundown of machinery is completed.

Plug and socket systems

Multi-pin plug and socket interlocks

Plug and socket interlocks are suitable where access is infrequent, e.g. for maintenance, tool changing/setting purposes. They are not suitable on gates used for loading/unloading components.

As in the case of limit switches, where spindle rundown exists or where it is likely that tool breakage or machine damage may be incurred if the guard is opened during cycle, a shotbolt machine is required in addition to the plug and socket. Fig. 3 illustrates a typical plug and socket system.

Direct access is available at each gate by disconnecting the plug from the socket. This isolates power to the machine parts which may be a source of danger to the operator. In cases of machine rundown, where a shot bolt is fitted, consideration needs to be given to incorporating suitable time delay

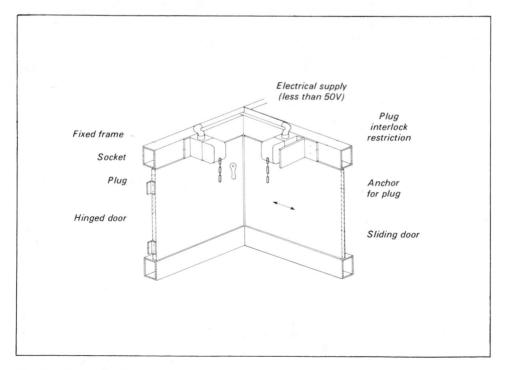

Fig. 3   Mechanical arrangement of plug and socket interlocks

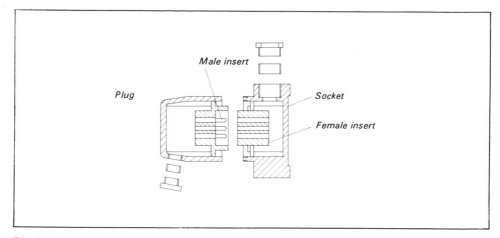

Fig. 4   Cross section of plug and socket

devices. To facilitate essential tool setting or fault-finding functions with guard doors open and power on the machine, a safe system of work must be established, e.g. the use of 'portable' deadman's controls. When operating with deadman controls, consideration must be given to stopping times.

Normally, for the purpose of interlocking machine guards, a six-pin plug and socket is used, as shown in Fig. 4. However, where the isolation of ancillary equipment is necessary before entering the machine enclosure, a larger multipin plug and socket may be used.

When replacing these, attention should be given to the position of the live connections in the plug and/or socket. This is particularly relevant when random pins are used.

The socket is used in the control circuit and mounted on the machine guard framework. The plug is securely mounted to the sliding or hinged door. To prevent the guard door being opened before the plug is detached from the socket, mechanical restraints may be fitted (see Fig. 3).

The electrical voltage should not exceed 50V between any conductor or any conductor and earth. To enhance the integrity of the system the electrical circuit incorporates the use of diode links or bridge rectifiers which respond only to direct current. The control relay cannot therefore close until the alternating current is rectified by passing through the diode or rectifier, which is housed in the plug side of the circuit.

To prevent the plug being engaged with the socket whilst the guard door is open, the length of chain or wire anchoring the plug to the door should be as short as possible. Positive means should be used for securing the anchor chain or wire to the guard door.

Although the system is sufficiently flexible to compensate for wear or slight damage to guards, connection or disconnection of plug to socket can be difficult and could result in damaged pins.

The initial cost of the system may be relatively high, particularly in the UK

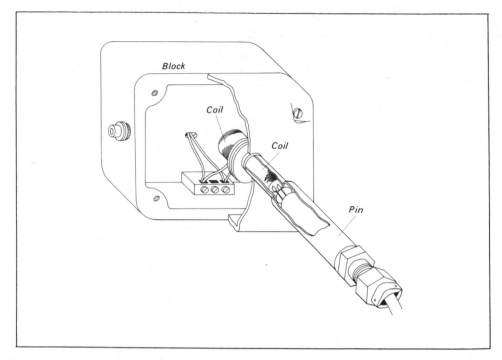

Fig. 5   Induction plug and socket

where specific circuit control features on high-risk applications are required by the Health and Safety Executive.

Induction plug and socket interlocks
This system of interlocking may be applied to the safeguarding of high-risk operations, where the multi-pin plug and socket system is prone to damage. They are suitable for use only where access to the inside of the guarding is infrequent, e.g. for maintenance, tool setting/changing purposes. They are not suitable on gates used for the loading and unloading of components or on machinery where rundown is experienced.

Immediate access is available through each interlocked gate by removing the pin from the socket which effectively isolates power to the machine parts that may be a source of danger to the operator.

Due to the simplicity of operation of the system, minimal maintenance of the component parts of the interlock is necessary.

The interlock system consists of a primary coil of a transformer (female block) mounted on the framework of a guard and a secondary coil (male pin) anchored to the guard door. When the door is closed the pin is inserted in the hole of the block, thus inducing a secondary voltage in the pin. Removal of the pin isolates the machine, i.e. the electronic control is designed such that any break or fault in the circuit stops the machine.

As in the case of the multi-pin plug and socket, to prevent the pin being engaged with the block whilst the guard door is open, the length of flexible

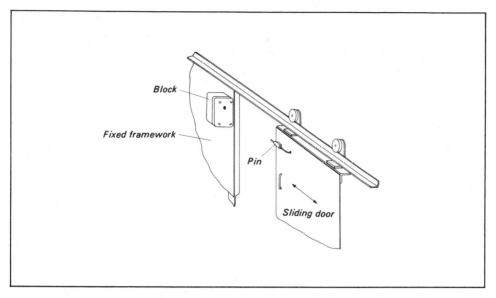

Fig. 6   Mechanical arrangement of induction plug and socket

conduit to the pin and the length of chain or wire anchoring the pin to the guard door should be as short as possible. Figs. 5 and 6 show a typical arrangement.

The induction plug and socket system is simple in construction, thus eliminating many of the problems experienced with the multi-pin plug and socket yet retaining flexibility in guard construction:

- Minimal effort is needed to disconnect pin from socket.
- Compact and less prone to damage and abuse.

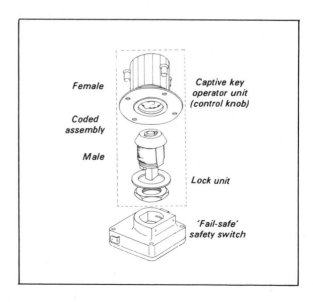

Fig. 7   Unimax switch

The initial cost of the system is high, although since it is less prone to damage, replacement is infrequent. For machinery with rundown, an additional shot-bolt arrangement would be necessary, thus increasing the cost of the total system above other alternatives. The system is suitable for use in most environments

Captive key systems

Unimax
These switches are only suitable where access through guarding is infrequent, e.g. for maintenance, tool setting/changing purposes. They are not suitable on gates used for loading/unloading components.

Direct access at each gate is controlled by turning the captive key knob to the open position. This isolates power to machine parts before access can be gained. Where machine rundown is experienced, a solenoid may be incorporated in the switch which prevents the captive key knob being turned from the 'off' to the 'open' position, until all machinery rundown is complete.

In the case of uniquely coded switches, when any part of the interlock unit becomes damaged or worn, the internal switch arrangement in both the male and female side of the switch will need to be replaced. In the case of common coded switches only the damaged/worn part need be replaced.

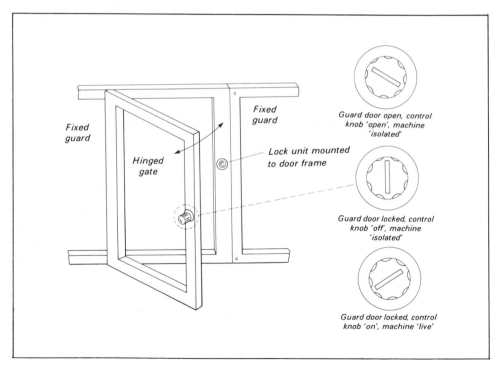

Fig. 8   Unimax three-position switching arrangement

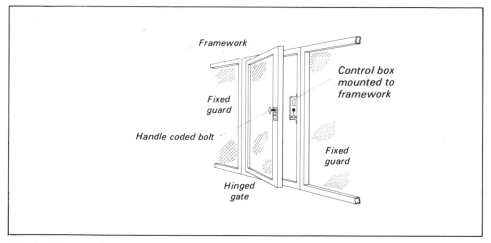

Fig. 9   Fortress interlocks

This type of interlock consists of three parts (Fig. 7). A tamper-proof switch housing a lock unit is secured to the framework of the machine guard and a captive key operator unit (control knob) which is permanently fixed to the guard door. This switching arrangement gives three conditions as shown in Fig. 8.

Replacement of component parts of the switch on-line results in lower maintenance cost. The initial cost of the system is relatively low. The system is fairly resistant to impact, dust, coolant, heat, etc.

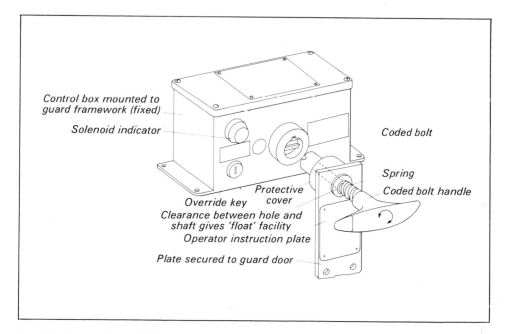

Fig. 10   Interlock unit (Fortress type 3119-018)

Fortress (Lowe and Fletcher) interlocks

These guard interlock units are suitable only where access through guarding is infrequent, e.g. for maintenance, tool setting/changing purposes. They are not suitable on gates used for the loading/unloading of components. These units are particularly efficient for the interlocking of machinery or equipment where spindle rundown is experienced and where coolant or oil may be a problem.

Access to the machine is available at each guard door by rotating the handle of the coded bolt on the gate side of the interlock anti-clockwise (Fig. 9). However, before turning the handle on the coded bolt, the machine controls must be switched from the 'auto' position to the 'off' position. This is necessary to release a solenoid in the interlock that traps the coded bolt securing the guard door. A light on the interlock unit indicates when the solenoid has released the coded bolt and the guard door may be opened.

Repairs of an electrical nature can usually be effected on-site, but mechanical failure of units due to wear or damage requires replacement of the unit.

The control box housing the solenoid and electrical contacts is mounted on the framework of the guard. The coded bolt is permanently fixed to the door of the guard. Provision is made for this bolt to 'float' in its anchorage, thus the problem of guard alignment when engaging the coded bolt is not so critical. The interlock is designed to facilitate the manual release of the solenoid should this become necessary due to a power failure or malfunction of the unit. An override key for this purpose is housed in the main electrical control panel (Fig. 10).

It is extremely difficult to defeat or override the system with the machine in a 'power-on' condition. In addition, the interlock control box requires rigid guard structure for mounting.

The cost of the system with solenoid and solenoid override facility is high. It is, however, suitable for use in most environments and particularly on wet machining operations where machinery rundown is experienced.

Photoelectric Guarding

Manufacturing Safety Coordination
Ford of Europe

The ever-increasing use of photoelectric cell safeguards throughout the manufacturing industry has given rise to the need for guidance on the installation and inspection requirements for the effective and reliable use of these systems. Based on the experience in plants where photoelectric systems have been in use for some time and with due regard to the legislation applicable, this paper attempts to give such guidance. In using a photoelectric machine guard, it is essential that the interrelated factors of beam setting, i.e. space detection capabilities, beam positioning and the machine stopping time are carefully considered.

With developments in technology and the need to remain competitive, the introduction of new and more efficient practices is essential. In the field of safety engineering the possibility exists to remove some of the restrictions associated with fencing/guarding machinery by the use of invisible safeguards commonly known as light barriers, light curtains or photoelectric cells. For some years photoelectric machine guards have been used on machines whose operation was cyclic in nature, using the system developed for use on press brakes. Due to the introduction and development of infrared techniques, it is now possible to produce systems which are suitable for use on many of the continuously operating machines.

Applications

Light barriers (hereafter referred to as PE cells) are defined as electro-sensitive means whereby an arrangement of photoelectric emitter and receiver devices can detect the presence of an object or person entering or present in a defined area. They may be used as:

- A trip device which stops moving machinery when the beam is broken.
- A presence sensing device to detect the presence of a person or object in the restricted area by interrupting the beam.

Depending on the restricted or danger area to be guarded, PE cells may be positioned vertically, horizontally, at an angle or in any combination of these modes. They may be installed on machinery where danger to persons will not arise from interruption of the beams, i.e. all dangerous movements are stopped before contact can be made with the dangerous parts.

Typical installations where PE cells have been successfully used are on automation-powered conveyors, presses, robot enclosures, and body framing bucks. Although in the majority of cases a photoelectric machine guard is used to prevent injury to persons by preventing dangerous motion of the machine when a person or part of a person is in a danger area, these systems may also be used in many other applications where their failsafe or self-checking facilities are required. These include restricting movement of a machine outside of its safe limits, e.g. a robot enclosure, the protection of expensive tools in moulding and stamping machines and detection of ejected parts from automatic machines.

Design principles

Two types of systems are commonly in use: the white light and infra red photo-optic systems.

White light (incandescent lamp)

An incandescent lamp has a considerable delay time before it is truly 'off'. This can affect the reaction time of the arrangement. Thus an incandescent or white light arrangement cannot be pulsed to a different frequency from ambient for practical purposes.

The incandescent lamp can be of the parallel beam arrangement or be diffused at the transmitter to give a large conical beam of light to the corresponding bank of photocell receivers. This arrangement caters for misalignment of the transmitter/receivers by as much as 30° and vibration problems met by parallel light beam arrangements. Some design arrangements overcome the problems of ambient light, but this should be carefully checked, along with the reaction time of the system relative to operators reaching the danger zone before it is returned to a safe state.

Infrared (light-emitting diode)

A light-emitting diode (LED) emits modulated light in the infrared range and can be of the parallel 'pulsed' arrangement, or the conical, constant wave modulated. The wave configuration shown on an oscilloscope would be the 'top hat' (castellated) configuration for a pulsed beam and a 'sine-wave' for the wave configuration. Each has unique advantages for specific guarding requirements, e.g. maintenance, misalignment, vibration, reaction time.

Infrared light screens are more commonly used to overcome the serious problems of sensitivity and ambient light. By using the LED as the infrared light source, it has the characteristic of being switched 'on' or 'off' instantaneously and at high speed.

System features

A PE cell has a transmitter and a receiver which are identical in form using a simple source in each unit and on opposite hand to each other. The light in each transmitter is diffused to give a large conical beam of light to the corresponding bank of photocell receivers. This feature gives rise to two cones of light in opposite form which together provides the required area of protection (Fig. 1).

In addition, a PE cell has a pulsed method of modulation. The cell comprises an optic head and a reflector unit (the optic head contains the transmitter and receiver) with its electronic controls (Fig. 2). An LED in the transmitter is fired, the light source strikes the reflector and is returned to the receiver. This firing sequence continues at an extremely high rate (approximately 280 times per second) and involves a considerable amount of gated circuit control on a synchronised basis. Some safety light curtains use a rotating mirror to direct an

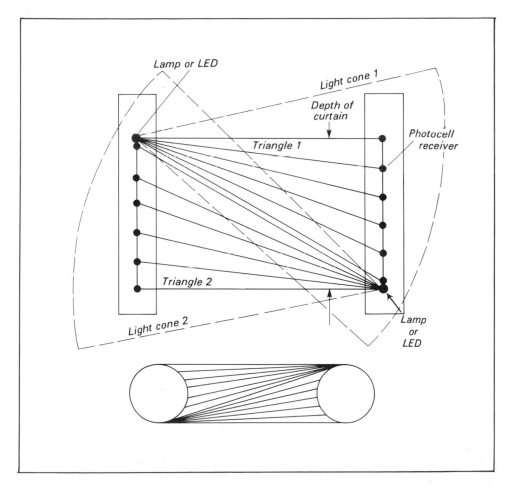

Fig. 1   Two light cones in opposite form provide the required area of protection

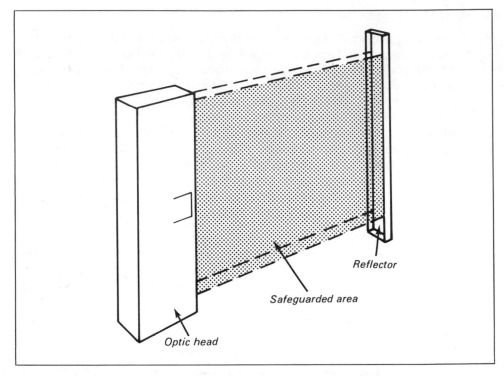

Fig. 2 Safety light curtain consisting of optic head and reflector

electronically modulated light in the infrared range, on to a parabolic mirror that causes the light to sweep the entire length and depth of the guarded area.

A photoelectric converter receives the emitted modulated light and produces corresponding electrical signals. Only the transmitted light frequency is evaluated by the photoelectric receiver. The signal pattern is processed by a logic circuit. As long as there is no interruption in the safeguarded area, the periodic scanning of the light beam produces a similarly periodic signal pattern. Each period lasts 10ms, if the light beam is interrupted the logic produces a 'machine stop' control signal through the relay output. This stop signal continues until a scan of the safeguarded area is completed with no interruption of the light beam. The optical principle is described further in Fig. 3.

Installation

When determining the position of the PE cell(s), consideration should be given to the distance it is sited relative to the dangerous part of the machine. It is essential that in all cases where PE cells are used as a trip device, that all dangerous movement has stopped before contact can be made with the moving part. Generally the formula for calculating this distance is:

$$D = v \times t$$

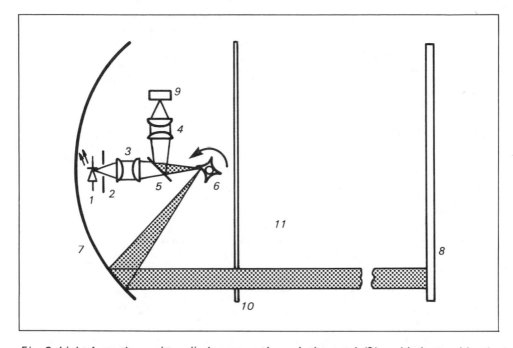

Fig. 3 Light from the emitter diode passes through the mask (2) and is focused by the emitter lens (3) through the beam splitter (5) onto the rotating mirror (6) where it reflected onto the parabolic mirror (7) and then through the window (10) to the reflector (8). The reflector (8) returns the light along the same path to the parabolic mirror (7) and rotating mirror (6) to the beam splitter (5) which reflects turning light to the receiver lens (4) where it is focused onto the photoelectric converter (9). The rotation of the mirror (6) causes the light to sweep along the entire length of the parabolic mirror (7), thus forming a continuous safeguarded area (11) between the mirror and the reflector (8). Because of the parabolic mirror, the light scanning the safeguarded area follows a path parallel to the optical axis. Each scan is preceded by an optically simulated interruption of the light beam (test pulse)

where D is the safety distance in millimetres, v is the reach speed (m/s), and t is the overall system stop time (ms).

PE cells should be installed and so arranged that it is not possible to stand between the cell and any dangerous parts of the machine. Where this is not achievable, alternative protective devices must be fitted to prevent this from occurring, e.g. physical obstruction.

When a PE cell has been penetrated by a person or object, the renewed movement of the machine should only be initiated by actuating a control unit. Also, cells that are not adversely affected by thermal or other radiation, noise, vibration, dust or excess water, should be used.

Examination, tests

Before commissioning machinery fitted with a PE cell guard, an examination and test must be carried out by a competent person. The results of this

- Check that the distance between the light barrier and the dangerous part is to the minimum dimensions specified on the information plate.
- Check that access to all dangerous parts is prevented from any direction not protected by the light barrier, e.g. by physical fencing or guarding.
- Check that the overall system response time is equal to or less than the time prescribed on the information plate.
- Check that it is not possible for a person to stand between the light barrier and the dangerous part.
- Check the photoelectric safety system for sensitivity according to suppliers' recommendations.
- Check the machine controls and connections to the photoelectric safety system to ensure that design requirements are met.
- Check that any muting arrangement (if fitted), meets the standard initially approved at the time of installation.
- Check the stopping efficiency of brakes, clutches, etc. as appropriate to the machine hazard.

Fig. 4 Schedule of examination and test

examination and test should be recorded and reports held on file by the user. Thereafter, an examination and test should be carried out every 6 or 12 months (Fig. 4).

A functional test, however, should be carried out by the user daily. This test should take into account and establish that:

- Access to all dangerous parts is prevented from any direction not protected by the PE cell(s), e.g. by physical fencing or guarding.
- It is not possible for a person to stand between the PE cell and the dangerous part and not interrupt the light.
- With the aid of the appropriate testpiece the PE cell is penetrated in at least three different places, i.e. close to each transmitter and in the centre. This test should be performed with the machine under power to ensure that all motion ceases on insertion of the testpiece.

Any external damage to the equipment or wiring, should be reported immediately. When determining the position of the PE cell relative to the machine, it is essential that the machine's stopping time is measured. Whilst this time can be measured using a suitable stop monitoring device, care must be taken to ensure that the device can be appropriately applied, set-up correctly when measuring, and that it is capable of being accurately calibrated.

Safety Mats

M.E.K. Graham
Herga Electric Ltd, UK

Robot guarding needs to be treated as an integrated system. No one guarding method by itself is likely to provide a total solution—nor will any one type of safety equipment be good for all applications. This paper is presented very much in this context, and provides information on one possible method, pressure sensitive mats. Safety mat design and construction, example applications, features, and costs are discussed.

Although robots are currently in the forefront of everybody's mind, there are no 'altogether-new' problems for the safety engineer to solve which he has not perhaps met in separate instances before. What may, however, be different in the case of adequately guarding robots, is that the combinations of hazard may be more extensive, the levels of risk sometimes higher, and the area to be guarded often larger. With increasing levels of automation come increasing risks in the workplace – both to the operator and the passer-by. For the operator – although he may no longer be closely tied to a particular piece of machinery in order to manufacture or process an item, his areas of risk expand in two directions:

- He may more often have a need to get right to the business end of the machine – to set, adjust, teach and maintain.
- He is no longer in *absolute* control of his machine's actions.

For the passer-by, the position is equally obvious. It is easy to become a casualty through ignorance and curiosity. Taken together with the fact that there will now be many cases where manufacturing companies are being rapidly exposed to dangerous processes which they have hitherto not experienced, there becomes an obvious need at the outset of a project to design-in a complete safety system as part of the total installation.

Robot guarding needs to be treated as an integrated system – where often some combination of special software provisions, fixed guards, perimeter interlocking, photo-electric devices and pressure sensitive mats will be necessary to secure an adequate level of safety. No one guarding method by itself is likely to provide a total solution – nor will any one type of safety equipment be good for all applications. Horses for courses!

Hence this paper is presented very much in this context, and hopefully will provide information on one possible method of guarding which can be acceptable in some applications, and which will most usually be relevant in conjunction with other safety features.

An introduction to pressure sensitive mats

Not everyone will be by any means familiar with safety mats and their widespread existing use in industry, and it may therefore be helpful initially to look at a few examples of their application.

The guarding of dangerous machinery with mechanical guards is not always possible or practical. In some processes, such guards can severely limit accessibility to the machine or to the product, and can often restrict visibility where this may be important. They can also reduce production rates, and it is not unknown for restrictions of this nature to even lead to their later removal where their use cannot be reconciled with the practicalities of manufacture.

One solution, therefore, is to use safety mats placed in dangerous access ways, which will switch off the machine when someone steps into those areas. By virtue of their simplicity and unobstructive operation which rarely

Table 1   Examples of industrial situations for the use of safety mats

Main observations	Additional points to note
Palletiser	
–free unobstructed path for movement of material	–other guarding (i.e. perimeter fencing)
–protection for personnel walking into the danger area	–can be very heavy duty application
Reelers and de-reelers	
–free access to machine to load/unload reel (Fig. 1)	–mat specially contoured to fit into precise areas
–mat acts as interlock to prevent loading arms operating	
Car assembly line	
–whilst operator is standing on the mat, the car cannot move forward (Fig. 2)	–very heavy and frequent pedestrian traffic
Tube bending machine	
–free movement of workpiece possible only be protecting 'shadow' area where personnel might venture (Fig. 3)	–on some types of machine is linked into microprocessor control to allow controller closer access during parts of the process
–no other form of guarding really practical	
–entirely flexible form of protection	
Coil compactor	
–in steel works, the machine compacts coils of steel wire and straps them automatically (Fig. 4)	–very heavy duty application
–fast access required for frequent adjustment	–harsh environment, hydraulic oil, etc.

interferes with the production process, they are now widely accepted by production management and machine operators alike.

Indeed, in some cases of general machine guarding, they are the *only* way to provide satisfactory protection in situations requiring: free movement of the machine or workpiece in areas where personnel might venture; and free access to the machine to load or unload material, to supervise a process or to carry out maintenance. Table 1 gives some examples, which serve to demonstrate safety mats most important characteristics: versatility, accessibility, reliability, suitable for industrial use, economic, and durability. It is these characteristics which are also of major importance in the design of safety systems for robots.

Safety mat design and construction

There are essentially two types of safety mat presently available; and whilst the primary result in both cases is the same, i.e. they stop machinery when someone steps on the mat, their methods of operation are significantly different.

One is air operated, the type manufactured by Herga Electric Ltd, and it is the type which will be specifically dealt with in this paper.

The second type is a Switchmat, in which a number of electrical contacts within the mat construction are operated when pressure is applied. Although this type were the original/earlier type of mat available, they do have specific drawbacks in wet or hazardous environments and may also be less suitable in

Fig. 1 Use of safety mats in reeler and de-reelers allowing free access to the machine for loading/unloading

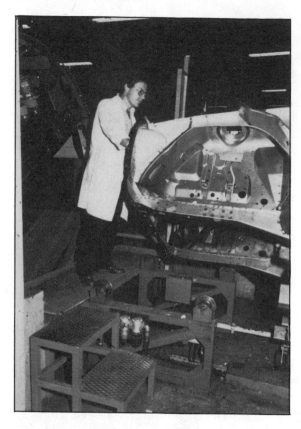

Fig. 2   Use of safety mats in car assembly line: whilst the operator is standing on the mat, the car cannot move forward

heavy duty applications involving heavy loads traversing the mat area or heavy items being dropped onto the mat surface.

The design of the air-operated type is simplicity itself. It is a unique system using closed-circuit blown air, in which an air blower in the control box circulates air through tubing in the mat.

The air pressure is measured at each end of the tube. Anyone stepping on the mat will squeeze the tube, causing the pressure levels at each end of the tube to change and operating two switches, either of which will switch off the machinery. These switches are in fact duplicated in each case to provide an acceptable level of electrical safety (e.g. in case of contact welding).

The control unit is relatively small (Fig. 5); it is the only part of the system containing the electrical switching, is in a sealed housing to IP65, and can be situated some distance away from the mat and operational area of most machines, if desired.

The mat construction is equally simple – based on a looped tubing system (usually of silicone rubber tube having remarkable resilient strength, yet is highly sensitive and has an excellent memory) fixed in a metal tray. A tough hard-wearing surface material approximately 12mm thick is placed on top – normally a cork/rubber composite although many variations are available for

Fig. 3 Use of safety mats in tube bending machines: free movement of workpiece is possible only by protecting 'shadow' area where personnel might venture

differing environments and requirements – and this is bonded into position with a flexible adhesive seal around the whole perimeter (Fig. 6).

On installation the complete mat unit is then fitted with a special aluminium extrusion which is used to fulfil various functions:

- When screwed down it fixes the mat firmly to the floor surface.
- Protects the edge of the mat from damage by falling objects.
- Clearly demarks a hazard area using yellow/black markings.
- Diminishes any tripping hazard by gently sloping the overall 20 mm mat height to floor surface level.
- Is used to provide a protective conduit for connecting pipes between mats when they are linked in series.

Special features of this mat system are:

- Since there are no electrical connections at floor level, mats are intrinsically safe in all environments and are especially suitable in wet and hazardous areas (e.g. flammable areas such as paint spray booths, and areas frequently washed down such as in the food industry).
- The mat's rugged construction with effectively no 'moving' parts, lends itself to reliable operation in a variety of arduous and heavy duty applications.

Fig. 4   Use of safety mats in a steel works coil compactor

- In spite of the above, mats have a high degree of sensitivity and provide relatively fast response times of the order of 150 ms.
- There are a wide range of mats available in size, shape and materials, which can satisfy most requirements.
- Large pressure sensitive areas are very easily created – and since the 'join-line' between any mats is also always pressure sensitive, these areas can be made up using a number of mat modules linked together in series air-loop connection.

Applications in robot installations

The only mat equipment generally suitable for industrial safety use will be of a type acceptable to the Health and Safety Executive (UK) and as such will usually be appropriate for normal risk applications in accordance with British Standard BS5304 (currently under review). Where safety mats meet the same requirements as those specified for photoelectric safety systems in Health and Safety Executive Guidance Note PM23, they may also be acceptable for some high risk applications. Often mats are more suited to a supporting (or secondary) role in the overall system.

Therefore, in relation to both the assessment of level of risk involved and to

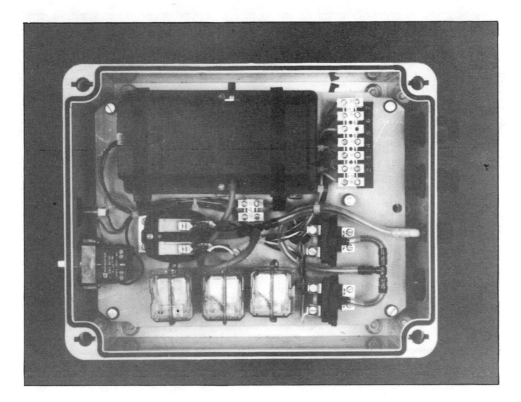

Fig. 5   Safety mat control unit

the final arrangement and function of a safety mat system, it will nearly always be relevant for the local Factory Inspector to comment on a proposed installation and to advise on its acceptability in that specific case.

A closer look now at some specific robot systems where safety mats are employed will highlight this and expand on some of the features mentioned previously.

Ford Motor Company

At the Ford Motor Company's truck factory at Langley in Berkshire (UK), four Unimate robot welding units are used on the production line to spot weld cab bodies together (see also page 267).

The primary guarding system to protect personnel from the machines is a steel safety cage enclosing the whole robot welding station and incorporating a key exchange system on all access points, which ensures that nobody can enter while the station is operational.

Authorised personnel can use the key system to gain access and power one robot at a time for necessary reprogramming under manual control. Automatic gates at the entrance and exit to the cage only open long enough to allow the finished cab out and the next cab in.

Pressure sensitive safety mats are fixed to the floor in this area, and whilst

Fig. 6   Complete mat unit

allowing free unimpeded movement to material on the production line, the mats will instantly shut down the whole welding station should any personnel enter the gates during the few seconds they are open. In this instance, safety mats are used as a secondary line of protection, but afford full accessibility and continuity of production.

Wavin Plastics
At Wavin Plastics in Brandon, Co. Durham (UK), another Unimate robot acts as a handling device for unloading large plastic mouldings (used in land drainage systems) from the moulding machine and placing them on an adjacent conveyor for transport to another part of the factory.

Here again, primary guarding in this area of the installation is achieved by fixed perimeter fencing totally enclosing the robot unloading station. Whilst a key interlocking system is used on access doors to the cage, if these were accidently closed when someone was inside, the robot could start operating automatically. A safety mat covering the whole floor area inside the cage prevents this possibility.

The large expanse and irregular shape around the robot, conveyor and cage perimeter makes this an ideal application for another type of mat arrangement. This is based on interlocking 300 m² ducktiles and incorporating the same air loop system and controls as previously described. Here, tiles can be cut on site to desired shapes to fit the floor area to be covered exactly, and can be quickly assembled to cover large areas economically (Fig. 7).

It is interesting to note that additional safety mats (of a heavy duty type) are used inside the moulding machine itself to prevent closure of the guard doors with anyone inside.

Once again, we have mats playing a secondary guarding role, allowing free movement of the machine and components, and demonstrating versatility and flexibility in the way they are installed and used.

Imhof-Bedco

Imhof-Bedco makes cases, cabinets, housings and enclosures for the electronics, telecommunications and computer industry, and has recently invested considerable sums of money in three paint-spraying robots with an overhead conveyor system and variable temperature controlled oven.

The RAMP robots are installed and operating at its Harpenden plant – two robots being used for spraying base coats and the third for the final textured finish.

The overhead conveyor passes into and out of each enclosed spray booth housing the robots. Access for both product and personnel is through the same apertures in the enclosure, and whilst a high level of safety had been created in the control system with strict supervision, it was additionally necessary to detect entry of any personnel via these access points into the restricted operational area.

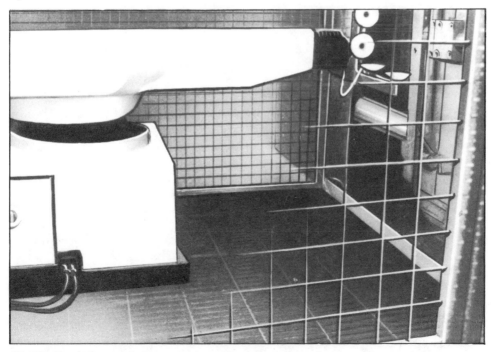

Fig. 7 Ducktile mat being used in conjunction with robot handling equipment

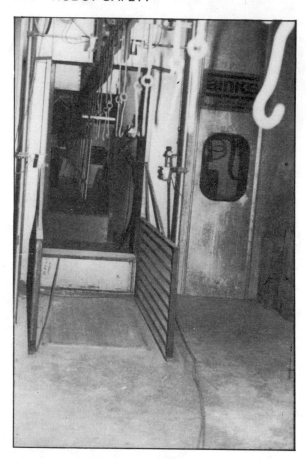

Fig. 8   Safety mat guards access to robot paint-spray booth

Safety mats, each approximately 1.75 × 1.0 m have been fixed to the floor in the walkways leading to these entry locations (Fig. 8), and with the additional provision of fixed side-fencing either side of each mat, it is impossible for anyone to enter the spraying zone other than by crossing the pressure sensitive area. Once again we can see how the uninterrupted flow of the production process can be maintained – but still providing adequate protection for personnel.

The project manager responsible for the planning and implementation of this installation, however, particularly stresses the importance of evaluating the full implications of safety requirements at the outset of the project and the need to design the necessary equipment into the scheme at its initial stages – and not add it on as an afterthought at a later date.

Duty, durability and reliability

It is clear from the three examples outlined above that the correct choice of mat for its proposed duty, environment and purpose is of paramount importance.

Only mats employing the appropriate features of toughness, durability, chemical resistance of the materials used in their construction, and fail-safe design, should be selected for the arduous and exacting conditions which quite often exist in areas around robot installations. After all, it is clear that one emerging feature of current interest in robots, is because of the possibility they create not so much to actually reduce manpower but to redeploy it away from the unpleasant working areas and harsher operating environments.

Not so for safety mats! They are increasingly being installed in these areas, and at ground level are subject to considerable battering both in load and chemical abuse.

Coupled with this is the fact that high levels of investment in sophisticated capital plant demand the highest possible return through more continuous and uninterrupted production. Reliability of all systems, including safety devices, is therefore extremely important.

Herga mat systems, based on the air-operated principles already described, employ a simplicity of design which limits the potential for unreliability. In addition, the many years of satisfactory performance in a number of those industries and applications into which robot automation is now appearing, shows clearly that long-term reliability – even in some of the roughest conditions – is certainly achievable. The robot is presenting no new environment problems for the safety mat! Failures in any piece of machinery must and do, from time-to-time, occur.

Of supreme importance in the failure of a safety mat system is the requirement that it should 'fail to safety' under all possible conditions.

With the air-operated mat system, any failure in the mats themselves, through tube disconnection, damage or blockage, results in shut-down of the machinery being guarded via the mat control unit.

The control unit itself has been designed to a high degree of electrical safety and provides dual circuit operation in areas of significant switches. For some models of control, additional relays are used so that the system continues to operate if any component fails. If the fault does not clear when the system is shut down, it cannot restart. Security relays with positive break contacts are used and the system cannot restart if the push-button becomes stuck down.

Costs

The cost of typical safety mat systems is, naturally, to a great extent a function of the area to be covered, the type of mat used and other specific features of each particular application. However, in broad terms an all-in figure of approximately £200 – £250 per square metre will give some idea of the cost involved for most heavy duty systems of the 5 – 10 square metre size. This is essentially equipment cost, but installation of any mat type is a relatively simple and quick affair and is, therefore, unlikely to add considerably to this figure.

By comparison with any other alternative – even when any feasible alternatives do exist – safety mats can generally be expected to offer a considerably cheaper solution for the same level of safety and protection, and can in many cases provide a number of additional operational benefits.

When viewed as a proportion of the overall capital cost of the robot installation, they are likely to be a notably small percentage of the total project cost.

Concluding remarks

This paper has endeavoured to show that the guarding problems for many conventional machines and in many existing industrial situations ally quite closely with a number of the hazards associated with developing robot installations. This is particularly true in respect of the major role which safety mats can play in the integrated safety system – that of secondary guarding.

Mats are especially relevant in protecting personnel in areas where unimpeded movement is necessary as part of a machine's manufacturing cycle – either for movement of the machine or of the product.

They allow free access to a machine for loading and unloading, for programming and maintenance – yet they can ensure that dangerous aspects of the equipment's operation are suitably isolated during this activity.

Safety mats are widely employed throughout various industries and are already demonstrating, over a number of years' usage, that they can provide long-term reliability and that they can resist harsh treatment and unpleasant environments.

There are various mat options available to suit different applications, including versions offering resistance to a wide range of chemicals and versions with the ability to withstand heavy loads. All such mat systems together with their control units should be fail-safe in all possible modes of failure.

The inclusion of pressure mats – as with all aspects of a total safety system – should ideally be reviewed in the early stages of the robot installation design, and such planned approaches will undoubtedly ensure a smoother start-up of operations with maximum levels of safety from the outset. However, whilst safety equipment manufacturers can give considerable information on the practicality and suitability of differing applications, the local Factory Inspector will ultimately determine a safety system's acceptability for the types of risk involved, and his advice should always be sought where relevant.

As to the future, there is undoubtedly a very strong undercurrent of development of new generations of robot device and new approaches to production automation. Exactly in which direction these will ultimately lead is perhaps not yet clear.

Perimeter Guarding

B. Powis
J.P. Udal Co.

The use of perimeter-style guarding has long been associated with the guarding of 'problem' machines; in the event of not being able to safely guard the machine with close fitting guarding the use of total perimeter guarding is adopted. Perimeter guarding construction and requirements for robot installations are discussed.

The general construction of perimeter fences usually consists of regular sized panels made up of a box section or angle section with either weldmesh or solid screen panels in between. The panels are usually fixed in place between suitably mounted pillars. This type of guarding allows for easy construction and layout of access doors into the machine working area. The height of the guarding is usually 2m from floor level, and can be taken right down to floor level or as in many cases finished short of floor level at approximately 150mm above the floor to allow for easy 'housekeeping' inside the perimeter guard area.

When this type of guarding is used on a robot installation careful consideration should be given to the actual application of the robot. A detailed risk assessment should be carried out to determine:

- The frequency of required access to the danger area.
- The method of working and suitable interlocking.
- The actual action of the robot within the enclosed area.
- Associated machinery involved with the robot and its application.

Perimeter-style guarding may consist of (Figs. 1–5):

- Fixed panel guards.
- Interlocked guards.
- Interlocked access points.

Fixed panels consist of solid construction panels, suitably fixed to either a framework or support. Fixings are 'security type' and require a deliberate act and use of tools to remove the panel from its position.

Interlocked guards consist of suitable guard screens operated from an open condition to a fully closed position before interlocks fitted to the guard screen

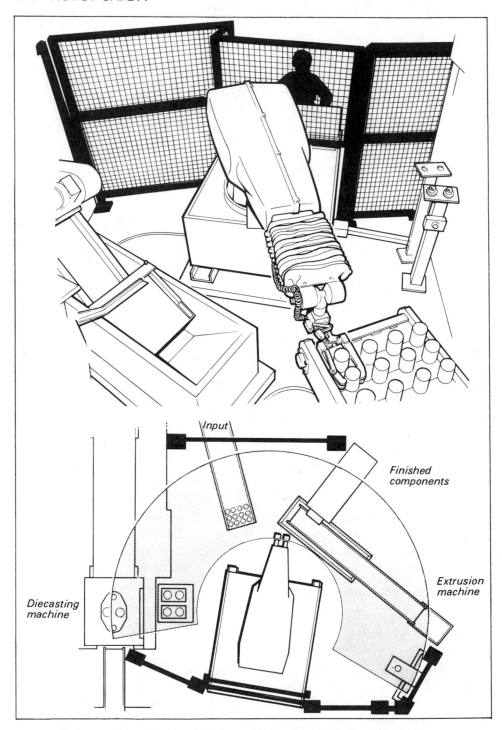

Fig. 1   Safeguarding of industrial robot with fixed and interlocked guards
(Courtesy MTTA)

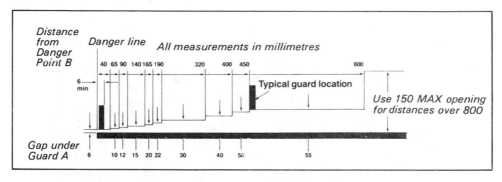

Fig. 2 Recommended openings in fixed guards showing the relationship between the gap A in a guard and its distance B from the danger point (Courtesy MTTA)

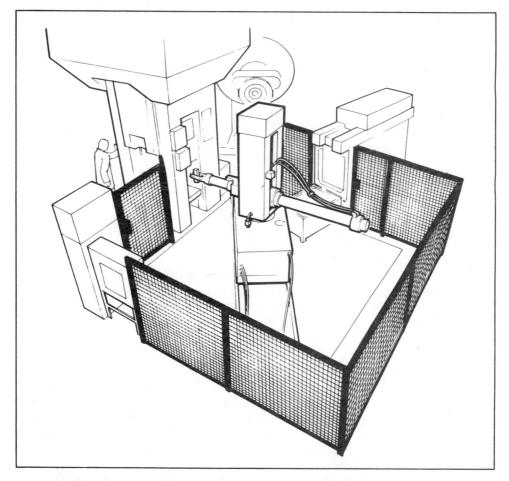

Fig. 3 Safeguarding of a robot with fixed and interlocked guards. The robot removes material from an induction furnace and loads an upsetting press (Courtesy MTTA)

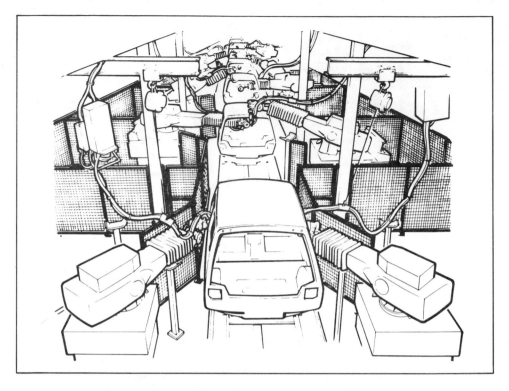

Fig. 4   Automatic robot welding line. Each robot is surrounded with safety
fencing of fixed and interlocked access guards. The fencing also prevents
access to the body track (Courtesy MTTA)

allow the machine to receive any operation signals. Interlocks are provided by
directly operated mechanical switch devices or suitable limit switch
arrangements (see also page 189).

Interlocked access points are of suitable door or access panels that allow
access into the operating area of the machine. The interlock fitted to the door
will prevent the machine from being operated while the door and interlocks are
in the disengaged condition. The interlocks consist of hinge-mounted limit
switches and cam switches or a captive key switch lock arrangement on the
door panel.

Positioning of the perimeter fencing should be such that it allows personnel
to move around with the access door open with sufficient clearance between
the extreme reach of the robot arm and the fixed fence. This is to prevent
trapping of a person between the robot arm and the fixed fencing. Sufficient
clearance should also be allowed for any other objects that may be situated
inside the fencing. Warning signs indicating that unauthorised access is not
allowed is highly recommended.

The interface of incoming or outgoing components, such as conveyor load
and unload systems, may require restricted access into the guarded area and
also need to be included with additional perimeter style guards.

Fig. 5   Loading of hobbing machine by industrial robot. Interlocked fencing prevents access to the robot or via the loading station (Courtesy MTTA)

Safety Sensor Systems

R. D. Kilmer
National Bureau of Standards, USA

With the increased use of industrial robots comes the concern for providing safe operating conditions so that accidents are avoided. One potential solution is to provide sensor systems which can detect intruders that enter the workstation and signal the robot control system so that an appropriate control action can be taken. This paper discusses the development of a prototype safety system using sensors. The characteristics of these sensors are discussed and a description of the application of this system to one of the National Bureau of Standards' (NBS) research robots is presented.

As the use of robots in industrial applications becomes more widespread, safety of both personnel and machinery, including the robot, is an increasingly important concern. Not only are there the obvious problems of injuries to personnel and damage to equipment, but also the problem of downtime associated with an accident. This downtime results in an increase of production costs, as well as a reduction of overall productivity – two key reasons for using industrial robots. Thus, the development of techniques for assuring safe operations is an important factor in applications of industrial robots.

Existing robot safety systems

Only a modest amount of work has been carried out on safety systems for industrial robots[1-6]. This is primarily due to the limited number and types of applications of industrial robots that currently exist and their relatively good safety record. The work on safety reported in this literature deals with four general aspects: personnel training, mechanical system design and workstation layout, control system hardware and software reliability and safety systems, including intruder sensing techniques. Work at the NBS has focused on the latter aspect, i.e. robot safety systems.

At present, the most commonly used technique for providing safety near industrial robots is to erect permanent barriers around the workstation, such as safety rails, fences or safety chains, in order to restrict personnel from entering these areas while the robot is operating. Although easy to implement, this approach is inadequate because no safety protection is provided for operators training the robot in teach mode or for personnel who are required to work nearby while the robot is operational. Also, the flexibility of the manufacturing facility is severely limited by erecting permanent barriers. As applications broaden to include more complex systems and cases where people may be working close-by, as in assembly operations, more sophisticated safety systems will be required.

One approach to solving this problem is to provide sensor systems which can detect intruders – personnel and other objects – that enter the robot workstation, and signal the robot control system so that an appropriate control action is made. In this respect two things need to be considered: the type of sensor system and the desired robot response.

Safety sensor systems

A variety of safety strategies can be developed regarding the type of intrusion and the desired response of the robot control system. There are cases when other equipment and personnel must be in the workstation while the robot is operational, e.g. during a routine maintenance check or during the operation of a robot in the teach mode. These situations obviously are to be treated differently from cases where someone, who should not be there, enters the workstation or some piece of equipment or hardware is left in the robot's working volume. To handle such a variety of possibilities, several distinct categories of sensor systems can be envisaged. For the NBS work on robot safety, these sensor systems have been broken-down into three levels based on the region of coverage and the associated safety strategy:

- Level I – perimeter penetration detection around the workstation.
- Level II – intruder detection within the workstation.
- Level III – intruder detection very near the robot (a 'safety skin').

Level I systems provide an indication of an intruder crossing the workstation boundary, but they do not necessarily provide any information regarding the location of the intruder within the workstation. The simplest safety strategy would be to halt all operations as soon as an intruder crosses the boundary. This, however, would severely restrict the flexibility of the workstation in much the same way a fence would. Another approach would be to use the Level I system to alert personnel that they are entering a robot workstation and that they should exercise extreme caution, or to provide a preliminary signal to the robot control system to activate and/or check the status of other safety sensors.

Level II systems provide detection in the region between the workstation perimeter and some point on or just inside the working volume of the robot. The actual boundaries of this region are dependent upon the workstation layout and the safety strategy being employed for a particular robot design and mode of operation. In some cases, it may be permissible for personnel to be inside the workstation and perhaps even inside a portion of the accessible working volume of the robot while the robot is operating. In others, it may be necessary to slow down or halt all robot movements as soon as an intruder gets within a specified distance of the robot.

These two possible strategies illustrate that there are a variety of approaches that can be taken in designing the Level II system, particularly in terms of the areas or zones of detection and the resulting robot control action. In the design of the safety sensor system, the general approach for sensing intruders and the overall safety strategy are obviously interrelated and are two of the key design factors. Some of the general approaches for handling intruder detection are:

- Detection in a limited number of zones where the probability of a collision occurring is the highest.
- Detection at any location within a specified area around the robot (exact location of the intruder may or may not be known).
- Intruder tracking through the workstation.

Similarly, the safety strategies can be grouped as follows:

- Complete shutdown of the robot as soon as an intruder is detected (either an application of the brakes, if so equipped, or a software stop).
- Limitation of the speed of the robot when an intruder is detected and activation of appropriate warning alarms.
- Instruction of the robot to perform other tasks in another zone until the intruder leaves.
- Instruction of the robot to take an alternative path to avoid a collision – obstacle avoidance.

These alternative approaches illustrate that there are many ways of designing the Level II safety system. Although the type of transducer and sensing system can be generalised, the final design of the safety system and robot control scheme will have to be specialised for different robot/workstation layouts and types of robot operations.

Level III systems provide detection within the robot working volume. This type of system, sometimes referred to as a 'safety skin', is required for cases where personnel must work close to the robot, such as during teach-mode operations. In such cases, the robot must be operational even though someone is within the working volume. The Level III system must be capable of sensing and avoiding an imminent collision between the robot and the operator in the event of some unexpected movement. Because the distance between the robot and the operator is much less in this case, the response time of the Level III safety system must be much shorter than for the Level I or II systems. These smaller separation distances also impose a requirement for finer distance resolving capabilities in the Level III system.

The concept of three levels of sensor systems does not require that the three systems operate exclusive of one another. In fact, some overlap of the regions of detection coverage is desirable, since this could be used to provide additional checks of intruder detection. These additional checks would help to limit the number of false detection indications. However, for cases such as teach-mode operations where the Level III system is the primary system, the Level II system would have to be in a standby mode. These and other constraints would be factored into the total safety sensor system design, which would be tailored to the particular robot application under consideration.

Types of detection sensor

There are a variety of sensing techniques currently used to detect intruders. The majority of the applications of these techniques have been used to provide security for commercial businesses, military bases, and, more recently, nuclear power generating stations. Security sensor systems can be categorised in much the same way as the Level I, II, and III breakdown of safety sensor systems described in the previous section. Although not standardised, the three general types of security intrusion detection systems are:

- Point, spot or object.
- Perimeter or penetration.
- Area, space or volumetric.

As the name implies, point systems are used to detect the presence of an intruder at only a single location. Perimeter systems are used to detect penetration across a specified boundary by an intruder. Area or space systems are used to detect intruders anywhere within a selection region defined by the field of operation of the particular sensor being employed. The system can be designed so that the selected region includes the entire volume of some enclosed space such as a room.

There are some obvious differences between the design criteria for security and safety sensor systems because the intended functional operation of these systems is not the same. In general, security systems are not required to provide information about the instantaneous location of an intruder, only that a particular point, boundary or space has been penetrated by an intruder. For efficient operation of the robot, the safety sensor system must provide additional information about the intruder's location in order to develop a system which will minimise the number of unnecessary shutdowns. Thus, not all security sensor systems are applicable to the design of robot safety systems.

Security sensor systems can be broken down into the following groups: switches (dry contact mechanical, magnetic or mercury), metallic foil, wire screens, trip wires, pressure sensitive mats, ribbons or wafers; acoustic sensors; motion sensors (ultrasonic, microwave or infrared); capacitance sensors; and vibration sensors[7]. Considering the design of a robot safety sensor system in terms of the requirements of Level I, II and III systems shows that many of these sensors are not applicable. For example, switch type devices might be used for Level I perimeter detection, but once triggered, they would have to be reset to be functional again. Another example pertains to the motion detection type devices. Although these devices perform well for security purposes, they could not be used in a robot workstation, because the motion of the robot would trigger the sensor.

Other limitations of some of these sensors are: susceptibility to environmental effects, such as temperature changes, extraneous noise or vibration, and dust or smoke; and to background signals from other sources.

Thus, only a limited number of commercially available security intrusion sensors are applicable to robot safety system design. Of the snsors listed above, only the pressure sensitive mats and the photoelectric sensors were considered suitable for use in developing a robot safety sensor system. In addition, various types of ultrasonic and infrared sensors (other than motion detection devices) were also evaluated for the safety system design.

Prototype safety system

A prototype safety system was designed for the Stanford Arm robot, set up in the NBS robotics laboratory. A general floor plan of the laboratory in which the robot is located is shown in Fig. 1. In this figure, the dotted or shaded area corresponds to the region that the robot can reach with the arm at maximum extension. The cross-checked area represents the region considered to be the robot workstation for this project. The outside edge of this area defines the boundaries as far as design of the safety system is concerned. The safety strategy employed in this initial system will permit personnel to be in the cross-checked area, but not in the shaded area (i.e. the working volume of the robot)

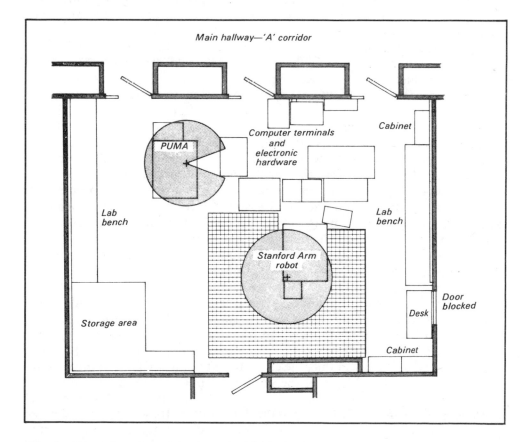

Fig. 1   Floor plan of the laboratory in which the Stanford Arm robot is located

while the robot is operating. When an intruder enters the cross-checked area, a warning alert (visual and/or audio) is broadcast. When an intruder is detected entering the working volume, the robot undergoes a software stop at its current position. The robot remains in this position until the intruder leaves the working volume. Once the intruder has left this area, the robot resumes its programmed task from the point at which is was stopped. For cases where personnel must be within the working volume, a hand-operated emergency stop switch is used to halt the robot.

The prototype safety system consists of both a Level I and a Level II sensor system. In addition, an emergency stop capability exists for use in situations where personnel conducting research must be within the working volume while the robot is operating. In such cases, the safety sensor systems are put on standby and a large, easily accessible, hand-operated emergency stop switch is used to halt the robot. Hitting the emergency stop button causes the electromagnetic brakes on each joint electric motor to be applied. Since this is not a software stop, the control software must be reinitialised before the robot can continue operation.

The Level I system is composed of a set of pressure-sensitive, industrial-grade floor mats positioned in the cross-checked area around the robot. When contact is made with one of these mats (5 psi), an electrical circuit is completed which is used to turn on a warning light. Pressure sensitive mats are used because they do not create any obstacles to isolate the workstation from the surrounding areas. Other types of sensors, such as beam-break photoelectric detectors, would have to be mounted on a stand or pole so that they would be at a reasonable height above the floor to detect an intruder. These mounting stands would act as a hindrance to movement around the perimeter of the workstation and might, in themselves, pose a safety problem because of personnel inadvertently bumping into them. Also, since these types of sensors only detect penetration of the workstation perimeter, they would have to be used in at least pairs at each boundary to be able to determine whether the intruder is entering or leaving the workstation. This illustrates another advantage of the mats in that the detection signal will remain on as long as the intruder is standing on the mats within the workstation.

In addition to the cross-checked area, another mat with a circular perimeter is located on the floor inside the working volume (i.e. the shaded area), except for the space occupied by the table and equipment rack, as shown by the rectangular outlines in Fig. 1. The perimeter of this mat corresponds to the perimeter of the region reachable by the robot arm. When an intruder steps on this mat, the resulting signal is used to stop the robot (a software stop), since the intruder is within the robot working volume. Fig. 2 shows the Stanford Arm robot mounted on a table, the circular mat in the area inside the reach of the robot arm, and a portion of the set of mats outside this inner circle in the region corresponding to the cross-checked area. Note that for this outer area, there are eleven mats. For this initial safety system, these mats are all wired together to give a single output signal. It is possible in future applications to look at the output of each mat to get an indication of an intruder's general

Fig. 2 View of the Stanford Arm robot showing the inner circular mat and a portion of the set of mats covering the general workstation area

location within this area. This information could be checked against the indications from other safety sensor systems to further substantiate the location of an intruder.

The Level II system used in this prototype design consists of an array of five ultrasonic echo-ranging sensors. These sensors consist of an electrostatic transceiver (i.e. transmitter and receiver) and the support electronics for determining the separation distance between the transceiver and some target. The basic operation of the sensor involves:

- Transmission of an ultrasonic pulse from the electrostatic transducer.
- Reception of any reflected signals using the transducer at the receiver.

- Measurement of the time-of-flight of the ultrasonic pulse from the transducer to a target and back to the transducer.
- Computation of the separation distance between the transducer and the target, based on the time-of-flight measurement.

In the prototype safety system, the ultrasonic echo-ranging sensors are used to determine whether an intruder gets closer than some predetermined minimum distance from the robot. If an intruder is within this distance, a signal is sent to the robot controller to halt the robot. To determine if there is an intruder present, a comparison is made between the measured transducer-to-target separation distance for each transducer (actually the time-of-flight is used for comparison) and a previously determined value, which corresponds to a point outside the reach of the robot arm. The reach of the robot arm when it is at full extension plus some margin of safety, which will permit the robot to be stopped before a collision can occur, determine this preset minimum distance.

Ultrasonic sensor characteristics

The ultrasonic echo-ranging sensors are similar to those used in a popular make of automatic focusing camera. The senors used in this prototype system were obtained commercially as part of an ultrasonic echo-ranging designer's kit. The sensor electronics were adjusted so that the transducer radiated an ultrasonic pulse or chirp between 7 and 8 times per second. This chirp, which has a duration of approximately 1.2 ms, consists of 56 cycles at four discrete frequencies: 8 cycles at 60 kHz, 8 cycles at 57 kHz, 16 cycles at 53 kHz, and 24 cycles at a nominal frequency of 50 kHz.

An important characteristic of the radiated pulse is its directionality. Fig. 3 shows a typical directivity plot for one of the electrostatic transducers. As noted from this plot, these transducers are very directional. The 10 dB down points are located at approximately ±10° off the centreline of the main forward lobe (±9° for this particular transducer). The two primary side lobes are located at approximately ±16°, with magnitudes between 13 and 15 dB less than the main forward lobe. These directional characteristics are advantageous, because with such a narrow forward beam and small side lobes it is possible to locate intruders more accurately and also to specify the areas of detection coverage best suited for a particular application. The disadvantage is that it takes several sensors to provide adequate detection coverage around the entire robot.

The sensor is designed to have an operating range of 0.9–35 ft (approx. 0.3–10.7 m). The limits of this range are a function of the timing requirements of the transmit/receive cycle and the distance attentuation of the ultrasonic pulse and its echo. The reliability of target detection by the sensor is optimised by three design features. The first feature is the use of a multifrequency pulse. By using a pulse with four frequency components, the probability of setting up a standing wave pattern between the target and sensor so that a null occurs at the sensor (i.e. effectively no echo signal at the sensor and, therefore, no detected target) is minimised.

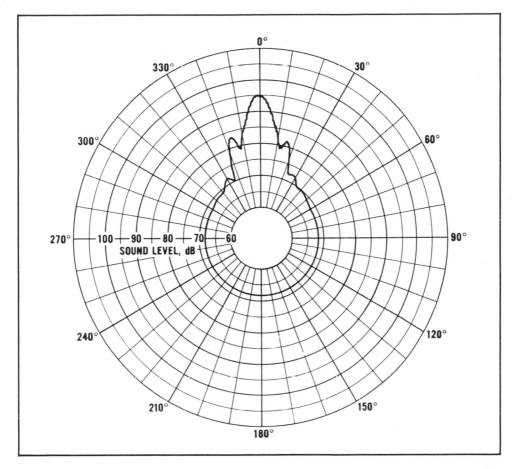

Fig. 3   Plot of the directional characteristics of the radiated ultrasonic sound waves from the electrostatic transducer

The second feature is the use of automatic gain control on the input amplifier to increase the signal-to-noise ratio of the received echo by increasing the amplifier gain. The reason for doing this is that the further the pulse and echo have to travel, the smaller the received signal will be due to distance attenuation. Since this distance is proportional to the time-of-flight, the gain increase is keyed to the system clock, which is also used to measure the time-of-flight. The gain is increased in finite steps at specified timing points until an echo is received.

The third feature is automatic bandwidth control of the input filter. This is necessary to minimise the effects of extraneous background noise sources. Initially, the bandwidth is relatively wide, so that all the frequency components of the pulse can be received. Again based on the system clock, the bandwidth of this filter is decreased to a constant narrow bandwidth centred at 50 kHz. The reason for this automatic bandwidth control is that at first background noise is not a problem, because the magnitude of the echo signal will be relatively large for close targets. For targets further from the sensor, the magnitude of the echo signal will significantly decrease as a result of wave spreading and distance attenuation so that background noise may become a problem. By narrowing the bandwidth to centre on 50kHz, the sensor can maximise the rejection of background noise and thus the possibility of false detection signals. The filter is designed to centre on 50 kHz rather than 60 kHz because the 50 kHz signal will be stronger, since the attenuation of sound waves in air is less for lower frequencies.

A number of tests were conducted to evaluate the detection capabilities of the sensor. A variety of targets made of different materials (e.g. plywood and polyurethane foam) were used, as well as human targets. In all tests, the sensor had no difficulties detecting the target regardless of the type of material. Also, it was determined that the sensor would detect the closest target and not necessarily the largest. This was determined by having a human target stand in front of a wall. In all cases, the human target was detected, which is exactly the result that would be desired when using this sensor in a safety system.

Operation of the prototype safety system

The prototype safety system consists of the pressure sensitive mats (shown in Fig. 2) and an array of five of the electrostatic transducers mounted on the Stanford Arm robot, as shown in Fig. 4. This prototype safety system is entirely hardware-based and does not have the capability of utilising joint position information or similar data from the robot controller. Because of this, it is necessary to mount the electrostatic transducers so that they move with the robot. This is done by fastening four of the transducers to a mounting bracket, which is bolted around a section of the support for the robot arm. In this position, the sensors rotate with the robot arm while still providing clearance for the robot arm to move up and down. The fifth transducer is mounted to the rear end of the boom which moves back and forth as the robot arm is withdrawn or extended.

The two primary areas requiring coverage are those around the gripper and around the end of the boom. Because these transducers have a ±10° cone in which an intruder can be detected, the positions of the transducers must be carefully selected to provide the desired coverage. One constraint to this is that the transducers cannot distinguish between an intruder, a workpiece on the table top, wire cables on the robot arm, or the robot grippers. Thus, the sensors must be positioned so that these other objects do not enter the operating cone of the five transducers.

With these design goals and constraints, the transducers were positioned to provide coverage in the areas illustrated in Fig. 5. The robot gripper and the end of the boom are the two areas of primary coverage, because these locations have the highest potential for a collision. There is no coverage to the sides. The only way the robot can strike an intruder is by rotating to the right or left. However, before a collision could occur, the intruder would be detected by one of the transducers, since the cone sweeps across the intruder's location before

Fig. 4   View of the Stanford Arm robot showing the locations of the five electrostatic transducers

the robot arm does. The other area not covered is directly in front of the robot gripper. Although this is not desirable and will be eliminated in more advanced safety system design, this is not a problem in this application, because the gripper is always operating over the table top, which extends beyond the reach of the robot arm.

The safety system electronics are designed such that any echo signal corresponding to a target (i.e. an intruder), beyond some minimum distance, are disregarded. Thus, as an intruder approaches the robot, an echo signal is received, but it is not until the intruder reaches this minimum distance that the signal is sent to the controller and the robot stopped. This minimum distance was set to be approximately 1 ft (0.3 m) beyond the reach of the robot arm, so that the robot can be stopped before the intruder takes one more step and can potentially be struck by the robot. This distance was found to be more than adequate to stop the robot and prevent a collision, even with the robot operating at maximum speed and the intruder walking directly towards the robot.

In operation, the safety system electronics are configured to give a single

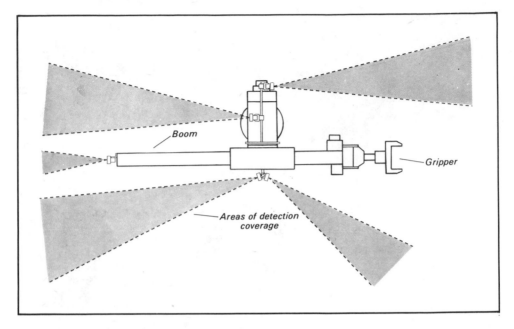

Fig. 5   Approximate locations of the electrostatic transducers on the robot arm and the areas of detection coverage

output to the robot controller when an intruder is detected by any of the sensors. (There is also another signal input, corresponding to the circular mat, which is independently sent to the controller.) When an intruder is detected within the prescribed minimum distance and the signal sent to the controller, the robot stays in this position. The minimum length of time the robot is stopped is 0.5 s (for a 25 ms sampling rate of the output of the safety system electronics by the robot controller). After the intruder has left, the robot continues to perform the operation it had started from the position in which it was halted.

One potential problem is false detection indications which unnecessarily cause the robot to be stopped. These false detection signals can be caused by electronic interference, extraneous noise sources with frequency components in the 50–60 kHz range, or echo signals from one transducer being received by another transducer. This was handled in the electronics of the prototype system by requiring that the intruder detection signal occur at least twice in a 0.8 s period. With an electrostatic transducer repetition rate between 7 and 8 pulses/s, this requires that two out of six consecutive pulses from the transducer result in an intruder detection indication. With this design, false detection indications were totally eliminated.

Concluding remarks

Overall, the prototype safety system performed well for all tests that were conducted. The pressure sensitive mats operated as designed and will be used in the development of future safety systems. For the electrostatic transducers, false detection signals, unnecessarily causing the robot to be halted, were totally eliminated. More importantly, the robot was capable of stopping within an adequate distance and in time to prevent any collisions with an approaching intruder.

Further development of this safety system utilising the ultrasonic sensors is underway. This involves incorporating a microprocessor in the system electronics so that information from the robot controller, such as joint positions, can be used to refine the operation of the system. These refinements permit techniques, such as difference mapping, to be utilised. Difference mapping involves storing a set of acceptable transducer outputs for various joint positions when no intruders are present. During operation, the safety system compares the currently measured values with the stored values for that particular set of joint positions to determine if an intruder is present. Utilising difference mapping permits sensors to be mounted on the robot near the gripper and at other locations not on the robot. This can be done because even if the robot or the workpiece triggers one of the sensors, the robot is not halted because the stored values would indicate that these conditions were acceptable for that set of joint positions. This provides better protection coverage without increasing unnecessary stops because of false detection indications.

References

1. Sugimoto, N. 1977. Safety engineering on industrial robots and their draft standard safety requirements. In, *Proc. 7th International Symposium on Industrial Robots,* 19–21 October 1977, Tokyo. Japan Industrial Robot Association, Tokyo.
2. Park, W. T. 1978. *Robot Safety Suggestions*, SRI Technical Note No. 159. SRI International, Menlo Park, CA, USA.
3. von Muldau, H. H. 1978. Safety at the workplace using industrial robots. In, *Proc. 8th International Symposium on Industrial Robots/4th CIRT,* 30 May–1 June 1978, Stuttgart. Society of Manufacturing Engineers, Dearborn, MI, USA.
4. Trouteaud, R. R. 1979. *Safety, Training and Maintenance: Their Influence on the Success of Your Robot Application*, SME Technical Paper No. MS79-778. Society of Manufacturing Engineers, Dearborn, MI, USA.
5. Woern, H. 1980. *Safety Equipment for Industrial Robots*, SME Technical Paper No. MS80-714. Society of Manufacturing Engineers, Dearborn, MI, USA.
6. Engelberger, J. F. 1980. *Robotics in Practice – Management and Applications of Industrial Robots*, pp. 89–91. Kogan Page, London.
7. Sher, A. H. and Stenbakken, G. N. 1979. *Selection and Application Guide to Commercial Intrusion Alarm Systems*, NBS Special Publication 480-14. National Bureau of Standards, Washington, DC, USA.

A Robot Safety and Collision Avoidance Controller

S. Derby, J. Graham and J. Meagher
Rensselaer Polytechnic Institute, USA

A study to assess the many sensor technologies as a means of creating a controller to achieve robot safety and collision avoidance has been conducted at Rensselaer Polytechnic Institute. Four sensor types were explored, and suitable candidates were tested in environmental tests and their performance on an industrial robot documented. These sensors have been integrated to form three-level detection zones, with several options of robot trajectory modification. The controlling algorithms on the monitoring computer have been established to facilitate the use of different combinations of sensors which may be application dependent.

The introduction of new automation into the manufacturing environment has historically caused worker injury. Equipment design, employee training programmes, safety rules, etc., eventually evolve to reduce the injury rate. Recent years have seen increased legislation to protect personnel in their workplace.

Industrial robots are being introduced into factories at an increasing rate. Three or four worker deaths have been reported for accidents involving robots[1] and there are expert predictions that US robot vendors could have sales of $2 billion in 1990[2]. The number of personnel interacting with these robots will necessarily increase as these robots must be installed, tested, taught, maintained, serviced and repaired. Anyone who has ever tried to program a robot to perform a complicated or precision task knows that close interaction with the robot is required. The increased number of robots installed coupled with more sophisticated applications could lead to some potentially unsafe and dangerous situations.

The present level of robotic safety consists of:

- perimeter warnings and interlocks,
- personnel training programmes, and
- preventative maintenance.

These three items are necessary for protection of workers, observers and intruders from injuries caused by a robot. Proper preventative maintenance will help to minimise the possibility of a run-away (loss-of-control) failure that could cause injury to equipment or personnel.

But a fourth item should be added. This would be a system that would use various sensors and a computer independent of the robot system to monitor the entire working envelope of the robot and take corrective action if unsafe conditions occur. Sugimoto and Kawaguchi's extensive robot safety study in Japan (see page 83) concludes that "Only when robots themselves are able to detect the approach of humans and perform appropriate actions to avoid accidents will safety in the human–robot workplace be assured."

Rensselaer Polytechnic Institute (RPI) is currently engaged in a research effort to investigate a stand-alone safety system that will monitor the area a robot can reach and take corrective action if predefined parameters are violated. This could serve as an additional layer of protection for many unsafe conditions that could occur in various robotic installations.

Classification of robot safety systems

To facilitate discussions of robot safety, zones or levels of protection have been designated by Kilmer (see page 223). With reference to Fig. 1 these levels are:

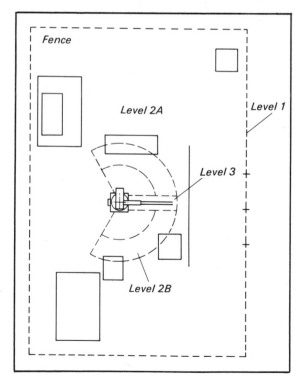

Fig. 1   Safety zones

Level 1 – Workstation perimeter penetration.
Level 2A – Area within the workstation but outside the reach of the robot.
Level 2B – Area within the workstation within the reach of the robot.
Level 3 – A small volume surrounding the robot arm which moves with the arm.

Level 1 protection is often obtained by the use of wire fencing which not only keeps out unauthorised personnel, but also protects personnel from flying projectiles if the robot should lose its grip on an object or a part should break into pieces. Other Level 1 safety devices which have been used or studied include pressure sensitive mats, photoelectric fences, and camera surveillance. Safety at Levels 2 and 3 has been much harder to implement and is the subject of this research. Work at the National Bureau of Standards using ultrasonic transceivers/transducers on a robotic arm for safety purposes has been carried out (see page 223).

A second classification scheme relates to how the sensor or sensors are used to improve safety. The two broad categories are:

● generic (task independent), and
● mapping (task dependent).

A generic sensor or system is one which is largely task or application independent. Thus the same sensor setup can be used on different robot installations, and for different tasks within an installation, with little modification. Most Level 1 safety technology is generic. In Levels 2 and 3, generic sensing usually means monitoring distances and velocities and comparing with some predetermined thresholds. An example is a sensor which monitors velocities and shuts down the arm if a runaway conditions occurs.

The converse of generic systems are task dependent systems which are best implemented by a mapping approach. In mapping, a teaching sequence is performed, during which the safety system records the information (i.e. distances or velocities) about the task being performed. During operation, the sensing system compares current information to the stored map information, and can signal a shut down if the deviation is too great.

Generic sensing is attractive because of its generality, ease of application, and the absence of a teaching step. However, certain situations and sensor systems are not suitable for a generic approach. A good example is trying to instrument the end effector so that it can pick up parts, but not injure humans!

Overview of the RPI safety system

As previously stated the purpose of the research was to investigate and develop a safety system for providing protection within the robot working envelope (Levels 2 and 3). This level of safety requires active real-time monitoring of sensory data. A number of sensing technologies were reviewed on the basis of sensing characteristics (range, ability to detect humans, etc.), durability and cost. From this list, four technologies emerged as having superior properties for robot safety, namely ultrasonics, microwave, infrared and capacitance.

For each technology, one or more representative sensing units has been obtained and tested. Commercially available units have been used where possible to help reduce the cost and to use proven circuitry. However, in many cases it was necessary to modify or redesign the commercial circuitry to allow computer control of sensitivity and other sensor functions.

To ensure the integrity and effectiveness of the safety system it is essential that the safety computer be constructed of high reliability components and incorporate self-checking diagnostics. It is also recommended that the safety computer be a separate entity from the robot control computer.

Safety system components

Ultrasound

Ultrasound sensing is based on producing a high frequency (above 20 kHz) sound wave, and then measuring the time interval until a reflection is detected back at the source. Thus it is necessary to both produce and detect ultrasound signals. The distance to the reflecting object is linearly related to the time delay by the speed of sound. Ultrasound sensing is used in intrusion detectors, for focus control in a popular instamatic camera, and for industrial gauging and ranging.

Two types of ultrasound sensors were obtained for testing: an electrostatic unit and a piezoelectric unit. The electrostatic unit used in the instamatic camera has a range of approx. 0.3–10.5 m, requires a 300 V supply, and costs about $100. This cost includes the analogue control electronics. The piezoelectric unit uses a lower supply voltage but is somewhat more expensive. Both units require additional circuitry in order to interface to the safety computer.

The ultrasound sensors performed quite well in both generic and mapped mode. However, there are a few problems which are the subject of ongoing research. The first deals with the relatively narrow detection beam of the ultrasound sensors (Fig. 2). Narrow beams are desirable for ranging, but for many safety applications a wider field of coverage is required. Horns and diffraction grating have been investigated to produce a more divergent beam, but more investigation is required on methods of attachment. A second problem concerns the relatively low repetition rate (10-20 Hz), covered by

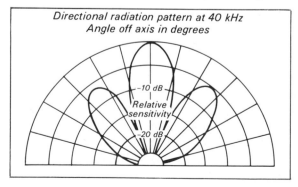

Fig. 2   Ultrasound lobe pattern

mechanical transients and circuit reset which must occur between pulses. This can cause problems with high speed robots which may move 12.5 cm in 0.1 seconds. This may be overcome by using multiple sensors with overlapping coverage.

Finally there are occasional false triggers, which are apparently caused by multiple echoes. A possible solution is to use multiple sensors which require multiple triggers before flagging a violation.

Microwave

Two types of microwave sensor are commercially available: presence sensing and velocity sensing. Microwave presence sensing uses an amplitude modulated signal but is very expensive ($2000 or more per unit). The microwave velocity sensor is commonly used for control of automatic doors and intrusion detection, and is moderately priced. These velocity sensing units are based on the Doppler principle that when a wave is reflected from a moving object, the frequency of the reflected wave is increased (or decreased) by an amount proportional to the object's speed.

The velocity sensing microwave unit obtained for testing produced an output pulse train proportional to the Doppler frequency shift. By measuring this frequency it was possible to determine the approximate velocities of

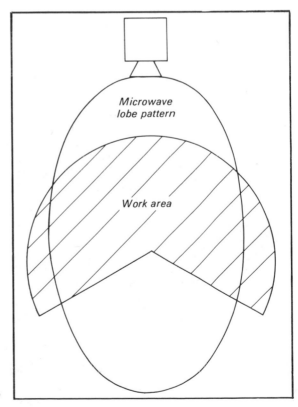

Fig. 3   Microwave lobe pattern

objects in the sensor's detection area, which is large enough to cover the working area of a T³ robot (Fig. 3). It is possible to isolate a microwave detector from outside interference by using a 6 mm wire mesh.

The microwave sensor was investigated for both the generic mode and mapping mode. In generic mode, the sensor is used to monitor the robot workspace and flags a violation if some velocity threshold is exceeded. This could be caused by an out-of-control robot or by an intruder, but either case warrants some corrective action.

In theory it should be possible to mask out the velocity components due to the robot, and this aspect was heavily investigated as part of this research. Although some success was obtained with mapping, there are three complicating factors. First, the microwave unit does not accurately measure very low velocities. The Doppler shift frequency is very low, and the pulse train output is erratic. This can be overcome to an extent, by simply disregarding small velocities. The second problem is that the sensor measures velocity components directly toward (or away) from the sensor, but not those tangential or lateral to the sensor. Thus an intruder walking laterally across the sensing area might not be detected. The third problem is that the reflected signal is directly proportional to the area of the reflecting object and inversely proportional to the distance from the sensor. Thus a human intruder might be masked by the signal reflected by a large robot such as the T³.

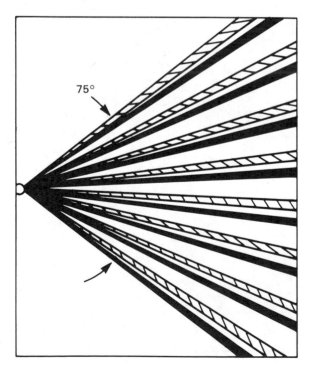

Fig. 4   Infrared fingers

Infrared

The basis for infrared sensor use is that all objects emit radiation if their temperature is above absolute zero. Humans emit radiation in a well-defined spectral range, so pyroelectric sensors have been developed that are tuned to the correct wavelength. These comercially available sensors are designed to detect intruders in an office or home environment. The intruder, moving in the surveillance zone, brings with him a temperature variation. This infrared radiation of about 10 μm is within the design range of these devices. Three types of these units were purchased in order to test them and to modify their circuit designs for computer interfacing.

In the pyroelectric sensor device, the change in temperature creates a change in polarisation, or the electrical charge. Thus, a temperature change will produce a current as a sensor output. This type of sensor device has a fast response and acts as a receiver to all human transmitters. But a simple pyroelectric device will change its output reading for a slow but steady room-temperature rise, otherwise a false trigger will occur.

The combination of two pyroelectric devices, often referred to as a 'dual detector', helps to solve the background sensitivity problem. Two similar but oppositely wired devices are lined out onto the same general work area, as are the fingers in Fig. 4. The first half is designed to produce a positive current when a temperature rise occurs, while the second half produces a negative current for a temperature rise. If the background temperature does change, the net effect of the pair is a zero current output. But an intruder, while in motion, will enter one of the zone's finger first and produce either a positive or negative current change.

Test results on the dual detectors for human intruders were very favourable in laboratory conditions. But other background considerations were also investigated. Standard lightbulbs produce radiation in the human range, leaving the sensors flooded with infrared signals. The motion of the robot itself was found to cause trouble. The Cincinnati Milacron T³ industrial robot produced radiation above, below, and within the human range while performing various tasks. So a threshold setting on the sensor's signal processing would not work. Even lower sensor mounting locations with their improved performance were not satisfactory. This meant that infrared sensors would not work in a generic fashion.

It was found necessary to use the infrared sensors in a mapping procedure. This would mask out the repeatable performance of the robot. Cyclic measurements were very repeatable, so an adaptive value mapping technique was used. The system was then able to detect human intrusions reliably. The adaptive value process takes sample readings during the first cycle for its reference values. The second cycle's readings are compared, and if they do not deviate more than a threshold level, the cycle is considered trouble free. The second cycle's values are substituted for the first cycle's values, thus becoming the new references. Thus, slow temperature fluctuations will be ignored as long as the background changes are less than the threshold values. The multiple fingers of the detection device can also be selectively masked out to keep a trouble source from constantly flooding the detection zone. The parabolic mirror for that finger can be physically masked to turn off that pair of fingers.

Capacitance

The capacitive sensor is currently used in the factory environment in order to protect the worker from injury with such machines as presses and stamping machines. Upon entry of the protection zone a relay is triggered in the device which is in turn used to control the motion of the machine. The device must only cause the machine to stop. The Occupational Safety and Health Authority will not permit such devices to restart these machines.

Three commercial capacitance, or RF presence sensing devices, were available at the time of this research[3]. Range of detection was one concern, since many commercial users are for limited range detection only. Since robot working environments often cover a large area, it is desirable to detect the intrusion several metres away since the robot, when at runaway speed, may reach 200 inches per second. The availability of schematic diagrams from the manufacturer also forced the researchers to eliminate one type of unit.

The capacitance sensing device is designed to protect the worker from injury caused by interaction with a moving machine. By determining the changes in the capacitance between the sensor device and the ground, signals are generated which can be used to control machinery. The sensing unit usually consists of a control unit, an antenna as the sensor, a coupler and the connecting cable.

The control unit contains the power and comparative circuitry and relays. The antenna can be constructed from copper, galvanised steel, or other conductive metals. The antenna shape is configured for the particular application. It is connected to the control unit with a coaxial cable. The antenna and ground act like two plates of a capacitor. The area of the plates and their dielectric constants have a direct effect on the capacitance. The comparative circuitry acts as an impedance bridge. With the fluctuation of the surrounding field of the antenna, one arm of the bridge is thrown from its balance condition to one where current flows through a galvanometer. As part of the device setup, the bridge is nulled to take into account the particular environment present. Any changes from this null will trigger the signal for the relay. Depending on the application, the relay can be used to sound alarms, cut off power, or apply a brake.

In applying the capacitive sensor to diverse robotic applications, the usually well-defined area of a static machine becomes one with many problems. For the antenna to be most effective, it should be placed close to the robot's end effector. This motion is quite different from the design's original purpose. The motion will take the antenna from its null position, possibly causing a false trigger. The motion of a runaway robot can be quite large, so the process of deciding whether or not to trigger must be made quickly.

A number of environmental factors have a known effect on the capacitive sensor. With a change of humidity or temperature the balance in the bridge circuitry will be thrown off, possibly giving a false trigger, since an air capacitor is being used as one arm of the bridge, any variation in temperature or humidity will change the dielectric constant of the air and thus the capacitance. The individual components of the bridge may also change if the control unit is allowed to face the environment.

The capacitive units all tended to drift with time. Many hours of no motion will cause a false trigger. For a mapping scheme to work, a constant nulling circuit was designed. This has helped to solve the problems, but further refinements may still be needed.

Integration of sensors via the safety computer

All of the individual sensing systems described above are available in commercial packages that operate in the threshold-alarm (generic) mode, particularly for intrusion detection. While such self-contained units can be used to a limited degree in a robot safety system (primarily at Level 1), it is our contention that an integrated system under the control of a central microcomputer is superior for the following reasons:

- Allows more sophisticated processing of signals than simple threshold detection.
- Can combine information from several sensors to make a shutdown decision.
- Filtering of false triggers is more easily done by a microprocessor.
- Allows tailoring to the particular application.

The development of the safety computer is proceeding in two steps: development of the interface hardware and control software on a microcomputer development system, and development of a stand-alone target system suitable for industrial application.

The National Semiconductor 6608 Development System which was used in this project features an 8086-based board computer, serial and parallel interfaces, a programmable interval timer, an analogue cord, floppy disk drives, and the CP/M86 operating system. This unit was chosen because the 8086 is widely used in industrial control applications. Most of the software was written in PASCAL.

The control software was structured into four menu-selected modes: system configuration mode, generic mode, teach mode, and run mode. In the system configuration mode, the user tailors the sensor system for a particular application by selecting sensors to be used, and by initialising control parameters for each sensor system. This mode can also be used to display stored data tables for mapped sensors. In generic mode, only sensors which are selected for generic operation are activated. This mode is an appropriate operational mode for robot teach operations, during which new tasks are programmed with the use of a teach pendant. As part of the teaching process, new data should be collected for the mapped sensors, and this is handled by the teach mode software. Finally, run mode activates both generic and mapped sensors to achieve a fully operational safety system.

Directions for further safety research and development

The robot safety research project attempted to address the various difficult and somewhat contradictory problems that exist in developing a system that would have acceptance on the factory floor. Any independent robot safety system must be introduced into the industrial environment in such a way as not to add an additional layer of false security to personnel. Additional personnel training, etc., will be required to facilitate the implementation.

The safety system itself, to be practical, must to a reasonable degree be able to satisfy the following criteria for any given application:

- inexpensive,
- fail-safe,
- reliable,
- highly immune to false triggers,
- non-fragile,
- capable of working in an industrial environment,
- easy to install,
- hard to bypass, and
- fairly simple to program.

A great deal of work remains to be done before such a system could be introduced on the factory floor.

Algorithms must be user-friendly, allowing sensor violation parameters and teaching methods to be utilised quickly and effectively. The modification of the safety systems program should not add unreasonably extra effort after a modification of robot sequences.

As at many industrial sites and universities, off-line programming capabilities exist at RPI for defining data points on a graphics terminal by moving a robot image through a simulated workspace. These points can be automatically down-loaded to the robot controller and then sequence the robot. With detailed descriptions of the workspace, software programs could be written that would generate the safety system program at the same time the graphics simulation generates the robot program.

Much research still needs to be done on robotic safety. RPI feels that research in this area is important and timely for the manufacturing community. We thank the companies financially sponsoring our research and RPI will continue its commitment to robotic safety research.

References

1. *First Human Killed by Robot.* United Press International Dispatch, July 1982.
2. Conegliaro, L. 1983. Trends in the robot industry (revisited): Where are we now? In, *Proc. 13th Int. Symp. on Industrial Robots and Robots 7,* Vol. 1, pp. 1.1–1.1. Society of Manufacturing Engineers, Dearborn, MI, USA.
3. Jacobs, C. 1983. *A Modified Capacitive Presence Sensing Device for Robot Safety.* M.S. Thesis. Rensselaer Polytechnic Institute, Troy, NY, USA.

5
Case Studies

Two of the case studies are by manufacturers of safety equipment the other three describe the users' installation. Together they serve to highlight the problems of robot safety and illustrate specific situations which have been chosen by the companies concerned in the light of the technology available at the time.

The first paper by Hugh Duffy describes the use of photoelectric guarding. The paper by Peter Burton describes how typical robot safety systems developed at British Leyland covering a range of different systems. In the third paper Danny Yule describes the use of Unimate robots for the welding of truck cabs at the Ford Langley plant where the coordination and programming of four robots to operate simultaneously raises some tricky safety problems during the programming phase. The fourth case study outlines a range of applications at Rolls Royce Ltd. Particularly interesting is the use of the machines themselves as part of the perimeter guarding. The paper by Brian Powis describes how perimeter guarding and associated equipment has been used in an application at Associated Steels and Tools, an associate company of J.P. Udal Co., the UK's largest manufacturers of machinery guards.

Erwin Sick

H.J. Duffy

Industrial robots have been used in Europe, primarily in the automotive industry, since about 1970. The early applications were mostly limited to single tasks and so the guarding problems were relatively easy to solve. Generally, the potential danger area created by the industrial robot was completely mechanically guarded. However, as the tasks performed by industrial robots have become more complex and as industrial robots have been integrated into complex production processes, mechanical guarding has been supplemented by non-contact guarding. Thus, non-contact guarding is often preferred for loading, maintenance and service stations, and it has proved to be effective for these tasks.

Non-contact guards

Typical mechanical guarding is provided by the use of fixed wire mesh surrounds, interlocks on doors, the use of pressure mats and so on. Non-contact guarding on the other hand uses means such as photoelectric guarding (light beams and light curtains), infrared sensors, etc. The general principles of photoelectric guarding are discussed elsewhere (see page 199).

The advantages offered by non-contacting guards over mechanical guards are summarised below:

- It is easier to load and unload parts, particularly large parts, at loading stations. It saves time and is ergonomically better.
- They can allow easier and faster access for repair and servicing.
- The lack of covers provides better visual contact.
- There are no moving parts, as found on large mechanical guards, covers, doors, etc.

Fig. 1 provides a useful summary of guarding possibilities. It can be seen that there are various types of non-contact guarding (NCG), but this paper deals primarily with optical systems and in particular presents the experience of Erwin Sick Optic-Electronic Ltd whose products have been extensively used in many industries, particularly in the automobile industry.

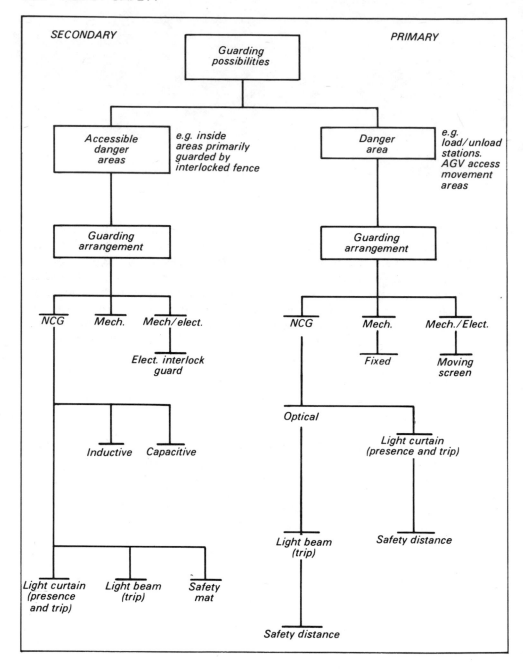

Fig. 1   Summary of guarding possibilities

Applications of optical devices

Typical areas of application of optical devices have been:

- The protection of personnel in paint shops where all necessary access and adjustment areas can be guarded by light beam switches.
- The protection of machinery can be obtained by using light beam switches when industrial robots are feeding parts into machine tools and presses.
- The protection of machinery in seam welding, spot welding and fettling of castings has been provided by light curtains.

The following are some of the requirements for guarding using light beams and light curtains: for accessible danger areas there is a need to guard all sides, there must be the required minimum distance between the light beams and the danger area, there should be immediate interrupt of the movement causing potential danger, there should be self-monitoring control of the guarded machine, the start operating panel should be outside the danger area and it must be possible to totally check (visually) the danger area from this location.

The use of photoelectric guards on robotic lines is now well established. Possibly the largest number installed occur on the £75 million Ford Transit line at Southampton and the Sierra line at Dagenham where in excess of 500 units are in use.

The case studies

Advice given in 'Safeguarding Industrial Robots'—Part I, issued by the MTTA in 1982 assumes that fixed or distance guards are the 'first-in-line' and refers to photoelectric guards under 'trip-devices' which, in general, may be used as secondary forms of guarding or where limited access is required, e.g. at load/unload stations. The other important consideration in the UK is Section 14 of the Factories Act 1961 which is discussed in general terms on page 3 et seq. The problem with perfect guarding as per Section 14 is that the removal of parts, loading of parts, maintaining, programming, setting and sequential operations must go on. Further comment is made on this later in the paper.

There are two situations considered in the case studies. The first is the protection of a worker who attempts to enter an area without valid reason or authority. The problem of guarding in this case is complex and essential, although the situation is of secondary risk. The second situation is the protection of an operator, e.g. at a load/unload area in which he is working legitimately. This is a high risk case but the problems of guarding are less complex.

It should be noted that effecting a cut out in a fixed guard to allow for a load/unload requires the area to be made good by an equally viable guarding system which, in all respects, must comply to the same evaluation factors as those which determined the fixed guard. The following describes the detail of photoelectric guarding systems used to protect operators in the two situations described above.

Case 1

Consider the movement of a product down a track as illustrated in Figs. 2 and 3. The track consists of robotic and manual workstations. A robot having performed its programmed movements will return to the start of the cycle and begin again. The product which has been worked on will be required to move on a carriage down a track to the next work area. The manual workstations can be manned by operators working singly or in teams. After a manned operation is completed the product will move on the track to the next operation which is assumed to be robotic.

The transfer of the carriage carrying the product into a hazardous area can prove very difficult if viewed in terms of Section 14 of the Factories Act 1961.

Fig. 2   Car body entering dip area

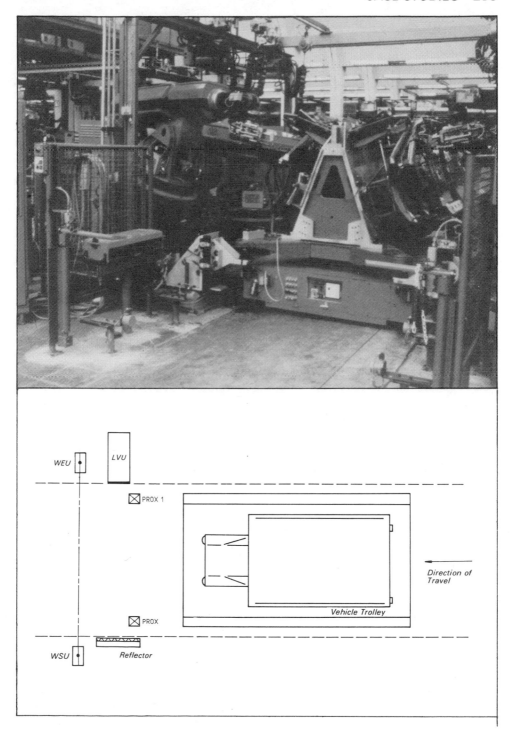

Fig. 3   AGV entering unmanned robot welding station

There is a need to allow access of a carriage or possibly an AGV to enter an area designated 'high risk' without hindrance to production and without personnel being put at risk. There is also the need to design out the possibility of system failure to danger. The fixed guard or interlocked guard must give way to the tangible photoelectric guard on the basis of practicality, but there is still a need to maintain a safety integrity. The following must be achieved:

- The carriage must be recognised and be allowed passage into the work bay.
- An operator must be recognised and all further hazardous movement within the work bay must cease.

To allow access of the carriage through the light curtain requires the curtain to the muted for sufficient time to cover the length of the product on the carriage.

The location of the carriage is determined by two proximity switches or photobeams, which supply an inhibit signal to the guard. The distance of the leading edge of the carriage from the light curtain must be as close as possible to ensure that an operator cannot be pushed or arrange to proceed the carriage through the muted guard.

The proximity switches are sensitive to the metal of the carriage and are located such that they pick up signals from both sides of the vehicle.

Passing beyond the muted light curtain, the leading edge of the carriage will break a single safety beam, located as close as possible to the light curtain on the inboard side. This beam, when remade upon the passing of the trailing edge of the load, will re-energise the curtain.

A typical depth of curtain protection is 700mm and located 400mm from the floor line, giving an all-in height of 1100mm. This can be raised to 1800mm using a larger light curtain, but experience has shown that 1100mm is considered as practicable.

The means of achieving these functions electrically is simple. However, achieving this safely and without causing a permanent mute condition and failure to danger, is entirely different to the functional requirement and needs considerable expertise.

The system shown in Fig. 4 and described below assumes that the light curtain and the safety light beam both comply to that part of BS 6491 1984 'Electrosensitive Safety Systems for Industrial Machines' relevant to the sensing unit and both are certified accordingly by a UK test house recognised by the Health and Safety Executive.

With reference to Fig. 4, proximity switch 1 and proximity switch 2 are not detecting. D1 and D2 are de-energised. The light curtain (LVU) runs up and is interrupted once to test the circuitry. The reset button is pressed and D23 energises. The curtain enters a green condition, D1 energises and D22 de-energises. The reset button is then released. The safety light beam is unobstructed and goes to green; D3 energises, D4 de-energises and brings in D5 which in turn brings in D9.

The x–x line is closed and as long as the reset button has not been jammed the machine can run. (Note, if an operator entered the curtain at this stage D21 and D22 would de-energise and the hazardous movement would stop. The

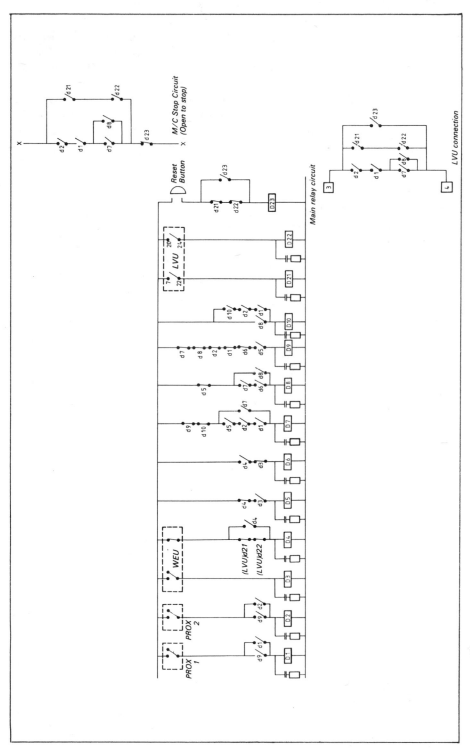

Fig. 4 Switching arrangement for the passage of a vehicle through the system (see text for details)

reset button would have to be pressed to restart.)

The carriage leading edge energises proximity switches 1 and 2 (inductive proximity detectors), and D21 and D22 energise, D9 de-energises and D7 can energise. This ensures that the x–x bypass line is now closed, i.e. contact D2, D1 and D7.

As the carriage proceeds it interrupts the curtain and a red condition results; D21 and D22 de-energise. The carriage proceeds further and interrupts the safety light beam which de-energises D3 and energises D4, thus bringing in D6. D5 now de-energises. D8 energises and holds over its own contact. D10 energises thus de-energising D7. When the trailing edge of the vehicle passes through the light curtain, D21 and D22 energise and the curtain is switched to the green condition.

The trailing edge of the vehicle then passes through the safety light beam switch (WEU) which in turn switches to the green condition; D3 energises and D4 de-energises. D6 then de-energises, D5 energises and drops out D8, enabling the x–x bypass line, i.e. contacts at D7 and D8 are open.

The trailing edge then passes over proximity switches 1 and 2 causing D1 and D2 to de-energise. This drops out D10 and the system returns to the start of the cycle.

If an interruption occurs during this cycle the light curtain will latch in the red condition until the reset button is pressed. All relays are checked for opening and closing during every cycle. If one fails the system will fail to safety. To ensure the integrity of the system no 'short-cuts' can be taken.

Case 2

Consider a load/unload station where the work process is carried out in two areas:

- The operator loads, in this case a wing inner, and the part is clamped (Fig. 5). The load table turns and a robot performs a welding process.
- The part is returned on the return table, unclamped automatically and the operator removes the part.

During the clamping and movement of both the table and the robot the operator must be in a safe position.

The light curtain must not operate as a secondary guard but as a primary guard, a presence sensing unit and as a trip device. It must achieve the same ends as the fixed guard fitted to the unload bay. In order to do so it is necessary to guard against the possibility of an operator or maintenance engineer stepping into the robot area while the bay is at rest and while the guard is in a presence sensing mode only. This is achieved by fitting a single (or double) safety beam, located 650mm high and 100mm inboard of the fence. Breaking the beam will result in the isolation of power to the hazardous movement. The beam is independent of the presence sensing guard.

The light curtain is located horizontally at approximately 750mm high, side fencing is 200mm from the floor level to 2m overall. It should be noted that the side fencing is located above the guard in such a position that the light curtain cannot be used as a step.

Fig. 5   *Operator loading a wing inner at a load/unload station*

A typical safety interruption line x–x is shown in Fig. 6. There are three external relays D1, D2 and D3 fitted with spark suppressors. The safety line x–x is formed using a series of contacts of these relays, i.e. normally open contactor D1, normally open contactor D2 and normally closed contactor D3. Safety lines x–x should be fitted in series with control lines which are responsible for actuating all dangerous movements.

The safety line should be fitted as electrically close as possible to the final switching solenoids. Interposing relays should normally be avoided, since such relays can remain mechanically 'stuck' in the energised position. If interposing relays are employed they must be checked during the machine cycle to ascertain correct operation. The current practice in the UK requires the use of

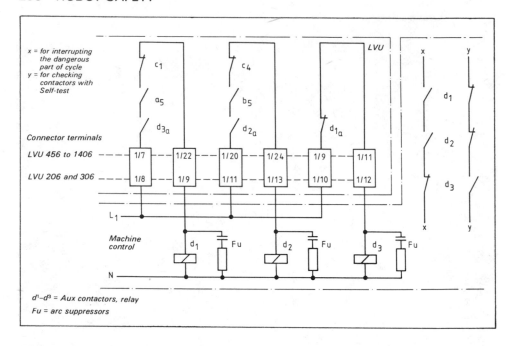

Fig. 6   Typical safety interruption line

two positively energised relays to form a safety line, thus the use of relays D1 and D2 should be considered mandatory. The use of relay D3 is optional.

When interfacing PLC devices generally, the following 'best practice' should be observed in order to overcome potential hazards, should such a device fail in an unpredictable manner. This subject is discussed in detail in the HSE Occasional Paper Series OP2, 'Microprocessors in Industry'.

Safety lines x–x may be used as an input to such devices, but only in the sense of a status input, indeed many such devices require a guard status input for correct operation. However, for the resource previously outlined, interruption safety lines should be implemented in all hazardous output control lines, again as near as possible to the final switching device.

The advice given so far is for interrupting all dangerous movements using safety x–x lines, it now remains to check the correct operation of the slaving relays D1 or D2 (D3). The concept here is that if relays D1 or D2 (D3) become mechanically stuck there will still be protection via the other relays operating normally. However, in order to avoid progressive failures, the first failure must be identified at the earliest opportunity.

The method for achieving this is straightforward. The manufacturer determines a part of the cycle of the machine which is considered to be non-hazadous, and the contact supplied for this within the light curtain circuitry is open circuited (by a cam switch or relevant relay contact) for a period equal to or exceeding 100ms. The effect of this is similar to that of an obstruction in the light curtain field, and if relays D1, D2 (D3) are correctly operated they will

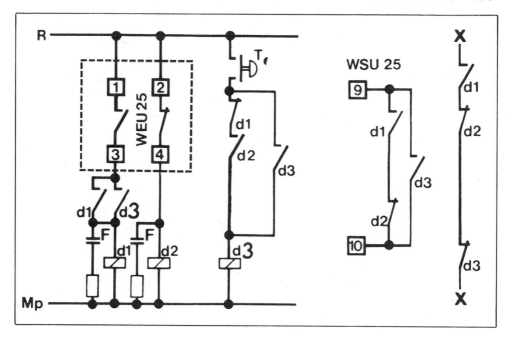

Fig. 7 Single beam configuration acting as a trip device only

change. The checking line y–y comes into operation and will normally be completed. This line should be placed in series with that part of the machine control circuit responsible for the return or non-hazardous movement; thus if a curtain controlled relay malfunctions the machine will be halted at a safe point in the cycle.

The single beam configuration, acting as a trip device only, is shown in Fig. 7.

It is vitally important that the reset facility is divorced from the actual control. An extraordinary act, i.e. the breaking of the trip circuit, demands extraordinary means to reset the system. It is advised that the reset button should be located in a visible position to the area entered and be key operated. VDUs, if fitted, should indicate the reset requirement which will cancel only when the normal circuit is made.

British Leyland

P. E. Burton

Robots have now been in use for some 20 years with the motor industry as the largest single user. To date no national or international standards have been developed to cover the specific requirements of robot installations. The UK, through the various bodies concerned with safety in industry, have taken a lead in producing guidelines to assist robot users and potential users. These guidelines are essentially based on automotive applications, but will be of value to all those concerned with the application of robots.

Industry as a whole in the UK has obviously taken steps to ensure safe operation in the meantime, since robots do not attract special consideration which makes them exempt from the Factories Act and the Health and Safety at Work Act. Different users have adopted different approaches to providing a safe robotic system. The following examples describe the steps taken by British Leyland in specific cases of robot application to ensure safe working under operational conditions – running, programming and servicing – so far as is reasonably practicable.

Five cases are described, showing how ideas developed as operating experience was gained by the company. This experience enabled the company to introduce the highly robotised and successful production lines which contributed to improving productivity levels up to the best of European performance.

General safety aspects

Automatic manufacturing systems can be potentially dangerous. It is therefore vital to provide protection for all personnel in the vicinity. The first, and most obvious, is to remove the manual element – e.g. a machine loader – from the area of danger. This can readily be achieved by locating the load position remote from the robot where the loading is done beyond the working envelope of the robot. Alternatively, physical limits to the robot's movement within its working envelope by means of 'goal-posts' on adjustable stops, can also be considered. It is important to consider the whole working envelope of the robot and not just that part which is used for a particular program; uncovenanted movement must be taken into account since a control malfunction could effect a movement to the extreme of any motion.

Casual observers or others working in the general area must also be protected, not only from the robot, but also from any other potential hazard associated with the robot or the task it is performing.

Programming presents an entirely different set of safety conditions. In this case the person 'teaching' the robot – the programmer – is the prime concern, but the possibility of others entering the area must be considered. Steps can be taken to overcome some of the problems presented to (and by) the programmer, but not all. What cannot be avoided with existing robots, controls and overall manufacturing systems, is the need for the programmer to be in close proximity to the robot during the teach phase. Observation of the completed program under some conditions also requires close proxmity during a robot run-through at normal operating speed.

Servicing, particularly 'live' servicing, can be performed under safe conditions when it is restricted to the robot control. If, however, observation of a suspected mechanical malfunction in a difficult or obscure location is necessary, it is not always possible to completely eliminate some element of risk.

Resistance spot welding

Spot welding is a point-to-point (ptp) operation requiring a large, powerful robot. Close proximity during programming is unavoidable.

The two examples given below are based on hydraulically powered robots. One application is an early standalone installation and the other a large, multiple application, with the robots operating as a 'family'.

Case 1 – Commission 1973

Here, a single Unimate 2000A robot is operating in conjunction with a rotary table welding a large subassembly. The robot is contained in a close mesh enclosure, approx. 2m high, with a single locked access gate. The component to be welded is manually placed on the rotary table which incorporates a 'job-in-position' switch and pneumatic clamping interlocked with the robot initiation. The system is initiated by means of a two-button safety interlock circuit. Various robot function switches are remoted for manual operation.

An emergency stop button is located adjacent to the load position and when depressed stops the complete system, including the removal of all electrical power to the robot. Opening the access door for maintenance or other purposes initiates a hold on the robot operation. An additional mushroom button is located adjacent to the home position support for use during welding gun maintenance, to ensure that access gate closure does not release the robot from the hold state and reinitiate the operational sequence.

During the teach phase, rotation of the table is inhibited to protect the programmer from uncovenanted movement.

Case 2 – Commission 1980

A large vehicle body finish welding line (re-spot line) incorporating 14 Unimate 2000/2100B robots, linked to an indexing transfer system. The line runs fully automatically with the robots forming the final stage of a complete body framing and welding system – earlier stages of the line consist of multiwelding stations and a gauging station.

The robots located either side of the transfer are mounted on raised platforms which incorporate a walkway with a short stairway access via a gate. A perimeter, waist high barrier is mounted on the platform. In addition, each robot is contained within an individual pen with barriers on all sides; including that side adjacent to the transfer system – in this case an inner gate.

Each robot pen has a gate which is not locked – opening the gate immediately puts the robot into a hold condition via the safety limit-switch. The switch is operated directly from the gate. A local hold is also remoted to each robot location, external to the pen.

Inside each pen a key exchange box allows access to the transfer system – removal of the key, which affords access via the inner gate, inhibits the transfer system. In addition, the hold can be overridden by means of a keyswitch as a deliberate act, to allow a robot to be recovered from the work area via the teach pendant, or operated independently for maintenance purposes.

Various levels of stop are incorporated in the system; all inhibit robot movement. Breakglass emergency stops will inhibit the transfer system and remove hydraulic power from the robots; the electrical supply to the pump is removed and the line pressure dumped back to the reservoir. Dumping is vital on a high capacity hydraulic system which is also supported by an accumulator, if movement is to be inhibited rapidly. Under this stop condition the robot control system remains energised to enable a program step to be retained.

At the end of each walkway mushroom headed emergency stops are fitted to enable all robots to be placed in a hold condition. Inside the pen each robot console has a mushroom headed emergency stop, also accessible at all times. Operation completely 'kills' the robot – hydraulics are dumped and all electrical power removed.

In addition to these operational/maintenance safety features, the robots are fitted with hydraulic flow restrictors which are operative (fail-safe) in the teach mode only. Any operation which requires the selection of teach mode, e.g. recovery from an emergency stop or reprogramming introduces the restrict element. All robot movement then takes place at a 'creep' speed.

Arc welding

Arc welding of light-engineering components, i.e. assemblies manufactured from sheet steel in the thickness range 0.8 – 2.0mm is best performed with robots that have ptp programming capability, with the ability to select speed between the points and the facility for linear or pseudo-linear interpolation.

The two examples given are based on small electric ASEA IRb6 robots.

Again, one example is a standalone installation and the other a large multiple installation with robots operating in family groups.

Case 3 – Commission 1980

A single ASEA 1Rb6 robot operating in conjunction with a rotary table, welding a small subassembly. The robot is contained within a 2m high sheet-steel enclosure containing tinted panels for viewing. The components are fed to the robot on a manually loaded four-position rotary table. At the load position a pressure-sensitive safety mat inhibits table movement until the loader steps off and presses the two-button, safety interlocked, sequence initiation circuit.

Access to the robot is via either one of two locked doors. A key exchange box is interlocked with the system electrical supply – turning the master key 'kills' the system and releases three captive keys. Two of the keys are removed; one opens the access door, the other is kept on the person.

The system can be overridden by a keyswitch for maintenance purposes by a deliberate act from within the enclosure.

Flexible screening between the robot workstation and the load station protect all personnel from the effects of uv light from the welding arc.

Case 4 – Commission 1982

A large multistation line with four groups of four ASEA IRb6 robots welding 2–2.5mm thick steel components. Two parallel manual load stations, protected by light guards and physical guards (press-type) initiated by push-buttons either side, feed the components on an indexing transfer to the first welding stations. The components then enter a small buffer prior to overhead transfer through two further welding stations.

The complete line is enclosed by 2m high sheet-steel screening with tinted viewing panels. Each group of four robots is contained within the enclosure, with local locked access and captive key interlock system as in Case 3. In this case, however, removing a key places all the robots in the group into a program hold state. Override is possible by means of keyswitches, however once one robot has been overridden the remaining three in the group are inhibited.

Due to the close proximity of the robots in any one group, servicing one whilst the others remain in operation is not possible. To allow servicing under running conditions would require extensive screening between robots to reduce the effects of arc 'glare' and uv light. Any such screening would restrict robot access if it were to be effective.

The complexity of the welding task makes close proximity to work zone essential during programming and subsequent trial runs of the operational robot program. Trials with closed-circuit television for remote veiwing of the welding areas proved abortive. A suitable location for the camera proved to be the major problem – even when one camera per robot was considered it was not possible to view large areas of the work zone. To assist programming safety, 'deadmans' handles will be fitted to each robot teach pendant.

Spraying

General Case

The operation of robots in paint-shop is usually confined to spray booths, therefore perimeter guarding is almost always complete, except for access doors. Where booths are not totally enclosed the perimeter can be effectively guarded by the addition of close mesh fences and access doors can be fitted with locks and/or interlocks.

Most automotive applications of robots to spraying operations are on moving conveyors. Programming is performed in real-time, requiring either the robot or a teaching aid (arm) to be used whilst carrying out the actual spraying operation. From a purely robot point of view this can be a completely safe operation; in the case of a teaching arm there is no motive power provided, just counter-balancing; in the case of direct robot programming the motive power can be switched off, again the system is counter-balanced either by springs or pneumatics, or a combination of both. When the robot application is on an overhead conveyor care must be taken to protect the programmer from a component inadvertently falling from the conveyor. This is not always easy to achieve, since the means of protection may restrict access to the work area.

All robots for use in paint shops must be intrinsically safe; for this reason they will almost certainly be hydraulically powered.

Handling

Component or assembly handling covers a wide range of applications, usually with the robot operating in the ptp mode.

Case 5 – Commission 1980

The following example is a complex task involving the movement of an assembly from a horizontal indexing conveyor to a low-level vertical indexing conveyor. This is followed by removing a similar component from the other end of the low-level vertical indexing conveyor. This is followed by removing a similar component from the other end of the low-level vertical conveyor loop – the component is now in reverse attitude – and placing it onto a high-level vertical conveyor. The whole operation is prioritised via a random program selection feature to ensure that the high-level conveyor takes precedence.

The robot is a Unimate 4000. The installation is contained within a 2m high meshed guard with two access doors. Located adjacent to one of the access doors is a key exchange box. The master key releases the access keys and removes the power from the robots via a remote stop which removes power from the hydraulic pump. Once inside the enclosure, maintenance personnel can override the remote stop for reprogramming or recovery from an in-program stop, as a deliberate act, with the access gate key.

Open access to the enclosure directly opposite the low-level vertical conveyor pick-up point is unavoidable since the robot is constrained by overhead obstructions – a vertical lift and retract is not possible. The open area is protected by pressure-sensitive mats connected in series with access key interlock.

When in the override state the safety mats and access gates are non-operative since programming, or any instance where the teach pendant is required, may need access to the open area immediately in front of the intermediate (low-level) pick-up point and movement through the access gates themselves.

Concluding remarks

Azimov's Laws of Robotics point the way, abstract they may be, but we all have an interest in safety. British Leyland, as with all robot users and potential users in the UK, look forward to the introduction of nationally accepted guidelines and standards. Only by these means can industry increase the use of robots in an orderly and increasingly safe manner to the benefit of all.

Ford Motor Company

A. F. J. Yule

The increasing use of automated equipment including robots throughout the motor industry, shifts the emphasis regarding safeguarding employees, from operators, who constitute the bulk of our employees, over to maintenance personnel.

The major difficulties in safeguarding maintenance personnel are associated with their function, which is varied, covering a range of trades and an inexhaustable number of tasks, as opposed to the traditional operator who has a specific function repeated endlessly throughout his working day.

Automated equipment is usually guarded by an enclosure to prevent access to moving parts and the truck cab finish welding station described here is no exception.

Maintenance personnel and programmers need to enter the confines of the station to carry out their functions, and as such a method of entry and a system of work is required for their protection.

Previous experience on equipment such as transfer machinery where access is necessary for maintenance, can be usefully applied to a robot station, particularly if previous experience includes the provision of inching or slow-running facilities. The movement of a robot through six axes under the direction of an electronic 'brain' is nevertheless unique and will give rise to situations never previously experienced.

It is important to know what is required to be worked on within the station, also to understand the type of work, and its likely effect, if effective safeguards are to be provided for maintenance personnel and programmers working within the confines of the station.

This case study outlines a system of entry into the confines of a robotic installation using a truck cab finish weld station at the Ford Motor Company's Langley plant as an example. Systems of work for maintenance and repairs are discussed as are the safeguards required for programming and testing.

System description

The finish welding station uses four hydraulically operated robots (Fig. 1). The cab is transported in and out on a conveyor, with a turntable sited in the centre

Fig. 1   The finish welding station

of the station. Each of the four robots has its own control console with the main programmable controller sited just outside the station. Entry into the station is required for maintenance, repair, programming and testing.

Hydraulic pressure of 950-1200psi is provided by a vane type pump driven by a 30hp electric motor with an unload valve acting automatically to maintain system pressure. A dump valve is provided to manually reduce system pressure to zero by allowing the fluid to be returned to the reservoir. The system of work had to contain instructions to open the dump valve before performing any maintenance operations on hydraulics and of course instructions to close the valve before returning the robot to operation.

Servo valves are provided on each axis to control speed and direction of movement. The direction of motion is determined by the direction called for by an electric signal, the magnitude of the signal determining how far the valve will open and therefore how fast the actuator will move. Failure of the servo valve can result in the robot movement on that particular axis continuing to its maximum distance. The system of work had to cater for this eventuality by providing some means of restricting each axis to a distance marginally greater than that of its required working area.

Three phase 440V 60-cycle electrical power is connected to the robot at the supply side of a fused main disconnect switch located in the main junction box.

Work on electrics must be carried out by qualified electricians with the equipment isolated.

Tests carried out on robots after repair are almost identical to the programming operation, therefore a single system of work was sufficient to cater for both operations.

Maintenance and repair

No one person covers all trades; therefore the method of entry and the system of work within the station had to cover electricians, plumbers millwrights, programmers, janitors and other personnel.

Maintenance and repair of equipment, e.g. clamps, conveyors, pressure weld guns and the turntable within the station, is similar to regular functions already carried out throughout the factory, and therefore the established systems of work apply. Power is shut off at the main console while the necessary work is performed.

The repair and maintenance requirements of the robot had to be considered in order to provide effective safeguards around which a practical system of work could be established.

Programming/testing

A Unimate robot is set up or programmed to work by driving it under manual control through the desired sequence of operation. This operation is also used to position the robot in any desired position for maintenance purposes.

Step-by-step successive point-to-point positioning of the robot arm and hand assembly is recorded in a digital electromagnetic memory. Also recorded are the timed waiting periods, the control of other equipment and responses to internal or external signals, which may be required in the program. The program, once recorded, will be repeated automatically and endlessly by the robot, until signalled to stop or shut-off.

Any object or individual blocking the path of a robot in operation, will be struck by the moving arm, resulting in personal injury, damage to the obstruction, or damage to the robot itself. Thus the system of work had to include keeping the working area clear of personnel and ensuring that obstacles would not be moved or placed in the working area after the robot had been programmed.

Teach control

Teaching is accomplished one step at a time, by moving the robot under the control of the switches on the teach control until the robot has been manipulated into the desired position.

The record button on the teach control is pressed to record the position in the memory via the position indicators (encoders) on the robot, making available to the memory system the location of the arm at that time.

In normal operation, the memory tells the arm where it should be and the arm is driven until the codes on the encoder matches the codes in the memory.

When codes match, the step is complete, and the robot goes to the next position indicated by the program. The process is repeated until the entire program is complete. An 'end of program' signal must be entered as the last step in the program. This final signal prevents the robot from continuing to operate on a previously recorded program which could have devastating results.

The system of work had to cater for this possibility. A new program must be tested step-by-step on automatic mode from the safety of the internal console. The system of work also had to cater for movement or replacement of encoders which will alter the programmed path of the robot.

Electronics

The design stage of this particular installation coincided with the period when safety engineers were bombarded with literature giving warning of the effect of lightning, static, radio waves, etc. on the software which would cause a variety of problems ranging from blank memories to berserk robots. It was therefore natural to assume that the reliability of electronics software had yet to be proven, and as such one should assume that malfunction was a regular feature to be endured.

The features designed into the station in conjunction with the system of work does not rely on electronics for protection. Automatic operations are carried out without the presence of personnel in the danger area, therefore hazard is non-existent albeit damage may occur. The system is shut down when personnel are required to enter the enclosed area.

Safeguards

Safeguards had to be designed into the installation based on the problems anticipated, around which safe systems of work and controlled access for authorised personnel could be provided.

Personnel had to be physically prevented from inadvertent entry into the station by means of 2m high open mesh perimeter fencing with automatic doors at the entrance and exit.

Pressure sensitive mats were placed on the floor inside the doors to cut power should entry be made by personnel when the doors opened.

A controlled access for authorised personnel had to be provided which would shut-off the power to the robots and associated equipment within the station prior to entry. A key exchange system was selected for this purpose. The entrance door key is held captive in the main control console and is released by inserting the master key into the console and turning, which cuts off all power to the robot installation. The master key is held captive in the 'off' position when the door key is removed. This cuts all power to the robot station. The door key can then be used to unlock the entrance door thereby gaining access into the enclosure and is held captive in the entrance door when in the unlocked position. The door is interlocked giving power only in the closed position making it impossible to remove the key without locking the door.

A facility had to be provided for powering robots for maintenance and programming purposes. Each robot has its own control console and a key to switch it on which was used in conjunction with the key exchange system. One key only was provided to fit all four internal control consoles. This key is held captive in the entrance door of the station, and released when the door key is inserted and turned to unlock the door. The door key cannot be turned from the unlocked position or removed until the robot key is replaced. There is one key for the robots, therefore one robot at a time can obtain a power supply with the key held captive in the 'on' position. No facility exists to provide power to the associated equipment when the station is occupied.

The robot's working area, or danger zone, had to be restricted to its minimum requirements. All axes were fitted with a combination of electrical sensors and mechanical stops. The electrical sensors prevent damage to the robot by stopping it before its movement is physically arrested by ultimate mechanical stops.

The internal controls of the four robots had to be sited in areas where the authorised person could operate the robot in safety, particularly during a malfunction. The robot contained within its mechanical stops would probably cause damage but it will not harm the authorised person who would hit the emergency stop button as soon as a malfunction became apparent.

Systems of work

The safeguards incorporated into the installations are the basis around which the system of work contained in Appendix A developed.

The master key to the key system is controlled by plant engineering personnel who restrict its issue to authorised persons only. Permits are issued to applicants on request by supervision, provided that the requisite training has been given. Authorised persons are reminded of the precautions contained in the equipment manual, extracts of which are contained in Appendix B.

Concluding remarks

The installation of a robot station for truck cab finish welding is unique and as such legal requirements or codes of practices are not as yet established.

Legal requirements where applicable were of course met, but rarely discussed. Protection of personnel from injury was considered a high priority at an early stage by design engineers.

The safety features, from which the system of work was developed, resulted from many hours of discussion between design and safety engineers, pooling their expertise to design out every possible hazard using features which were mutually acceptable, reliable and fail safe. This objective was achieved without significantly affecting the cost of the installation and having no effect whatsoever on production capacity.

Appendix A – Safe system of work

- Because of the inherent danger to personnel required to work within the arcs of self-acting robotic machinery, the following system has been designed and installed to provide a safe working environment. The four Unimate robots are isolated from personnel by:

(a) Total enclosure by a security fence at least 2 m high.
(b) A key exchange system with integral interlocks on the entrance gate to the enclosure. This system ensures that access can only be gained when the power supplies have been broken; the master key being held by 'plant engineering' for issue to authorised personnel only.
(c) A 'permit to work' system to ensure that only personnel who have been trained in robot maintenance and who hold an operators permit (Form 1609) will be allowed to gain access into the enclosure. The permit will be issued by the Plant Safety Engineer, following satisfactory completion of Form 30306, raised by the supervision of the person concerned.

- In order to gain access to the robot enclosure, the master key must be drawn from 'plant engineering' (see (b) and (c) above).

(a) The master key (A) is inserted in the control console and turned; this cuts all power to the enclosure and releases key (B).
(b) When key (B) is inserted in the gate and turned, it unlocks the access gate and releases key (C).
(c) Should it be necessary to activate a robot whilst working inside the compound, this can be achieved by closing the access gate and inserting key (C) into the individual robot's console.

Should the access gate be re-opened while the robot is still live, the power is interrupted, alerting the occupant to the presence of another person.

- On every occasion when work is to be carried out within the robot enclosure the master key must be drawn from plant engineering by the holder of a valid operators permit (Form 1609).
The permit holder, who will be fully conversant with the safety rules contained in Unimate 4000 Equipment Manual 398E, must check personally that the area and the robots are immobilised before allowing a non-permit holder to enter the compound to carry out his work.
Non-permit holders should be instructed to report completion of their work to enable the permit holder to check that the compound is clear and safe before locking up and re-activating the robots.
Should a non-permit holder, i.e. tip dresser, millwright, electrician, etc., working on other equipment within the enclosure have a requirement to have equipment moved by power, the movement is to be controlled by the permit holder who should take all necessary precautions to ensure the safety of all employees.
Persons under training to become permit holders must not be allowed to work within the compound on live equipment unless they are in possession of a temporary operators permit (Form 1609A), and are supervised/accompanied by a permit holder at all times. (Note, a master key should *not* be issued to a temporary operators permit (Form 1609A) holder.)

- DO take great care of keys, lock all doors and replace keys in each point, and return master key to plant engineering when your task is complete or prior to the end of your shift (whichever is earlier).
DO keep all disassembled parts spotlessly clean – a speck of dirt in the system can cause a robot to malfunction.

DO be cautious at all times – robots cannot think – one wrong or incorrectly entered program instruction can cause serious damage/injury.

- DO NOT take risks or chances.
  DO NOT guess – if you don't know ask your supervisor.

Appendix B – Precautions (extracts of Unimate robot equipment manual)

- DO NOT place yourself or mobile equipment in the area that the Unimate can reach when electrical power, hydraulic pressure, or pneumatic pressure is present.
- DO NOT perform electrical or mechanical maintenance of service on the Unimate until all electrical power is removed, and all hydraulic and pneumatic pressure is brought to zero.
- DO NOT open any hydraulic component or disengage any mechanical part of a motion drive train before ensuring that the motion will not move when acted upon by gravity.
- DO NOT change the position of the following switches, while the UNIMATE is moving (i.e. with the 'cycle start' switch on and the 'hold/run' switch in run): the the Accuracy 3, 'slow speed', or the 'speed adjust' potentionmeter setting is
- DO NOT teach a step with an actuator mechanically bottomed.
- DO NOT teach a direct reversal of a motion in Accuracy 3.
- DO NOT teach the 'end of program' step in Accuracy 3.
- DO realise that the UNIMATE will travel the shortest path from where it is, to where it is indexed to go (counter display).
- DO realise the path that the Unimate travels will be altered if either the Accuracy 2, the Accuracy 3, 'slow speed', or the 'speed adjust' potentiometer setting is changed.
- DO realise that adjustment, displacement, or replacement of an encoder, its coupling, or any mechanical drive component will change the path and positioning of that motion.
- DO realise that the Unimate cannot see nor reason. Any obstacle placed in its path of operation, after it has been taught a program, will be struck.
- DO realise that non-movement periods caused by time delay's, wait external circuits, a master controller, or computer interface make it unsafe to assume that the Unimate is not about to move because it is not in motion.
- DO realise that the Unimate can, during a failure mode, move to locations outside of its programmed area.
- DO consider what action is expected from the Unimate before a switch is toggled or button is pushed; then be prepared to stop the Unimate if a deviation is observed.
- DO disconnect the tape cassette recorder, tester, program editor, and teach control before placing the Unimate in production.
- DO realise that the only safe restriction to the movement of the Unimate is a mechanical stop.
- Do install the Unimate using an approved electrical installation standard (e.g. JIC).
- DO contain electrical or oil class fires using an approved fire extinguisher.

It is *not* normally safe to perform maintenance or service when within the reach of the Unimate until all of the warnings and procedures listed below have been adhered to.

- DO NOT perform electrical or mechanical maintenance of service on the Unimate until all electrical power is removed, and all hydraulic and pneumatic pressure is brought to zero.
- DO NOT open any hydraulic component or disengage any mechanical part of a motion drive train before ensuring that the motion will not move when acted upon by gravity.
- DO realise that adjustment, displacement, or replacement of an encoder, its coupling, or any mechanical drive component will change the path and positioning of that motion.
- DO place boom in horizontal position and install 'down support tool' 106 BG2. In creep speed, drive down-up motion so that boom just makes contact with the tool. If tool is not available, drive boom to full down position.
- DO remove air source from the Unimate and bleed entrapped air by operating the teach control 'clamp' switch(es). Verify zero air pressure by observing that the air pressure gauge has fallen to zero and the fingers, hand, tools, etc., are no longer able to perform their functions.
- DO depress the 'stop' switch on the control panel and verify that the green power on indicator changes from the 'on' state to the 'off' state.
- DO place the disconnect switch on the junction box to the 'off' position to remove electrical power from the Unimate electronic console and electric motor. Verify power has been removed by observing the amber line indicator on the control panel changing from the 'on' state to the 'off' state.
- DO open hydraulic gauge shut off and note oil pressure reading; then open dump valve. Verify absence of hydraulic pressure gauge *now* reading zero. Close dump valve finger tight.

Rolls Royce Ltd

M. C. Bonney

This paper describes robotic cell safety at Rolls-Royce Ltd through the eyes of a visitor. Rolls-Royce Ltd has now installed several robotic workstations as part of its developing Advanced Manufacturing Technology programme and the working enviroments create a favourable impression.

The level of robot safety awareness is high. One indicator was the availability of safety information. This included the MTTA Guidelines, and the paper on Robot Safety and the Law (together with the HSE checklist) which appears on page 3. In addition, Rolls-Royce Ltd uses a framework of standard safety. features which has been derived from its own experience and which is used if appropriate. Typically, this safety framework may be summarised as working with a full awareness of statutory obligations.

Guarding is by means of 2m high mesh enclosure. Where possible, the machines are used as part of the barrier. Conveyors enter through windows in the mesh barriers. Generally, there would be hardware interlocks located on the tops of the doors, strategically placed stop-buttons on the robot PLC and other locations such as by conveyors, guarding on conveyors, notices as appropriate, appropriate training, and a two-key system for access (one related to the PLC and the other chained to the cell door). The PLC key is withdrawn and this isolates the PLC before the door key may be used.

That being said, there is no rigid 'policy' and the safety and technical aspects of each cell are the subject of considerable planning effort, followed by discussions and negotiations with the appropriate safety officials and unions.

One possibility considered was that, where appropriate, new robot cells would have perimeter guarding 2m high with the top 0.5m constructed of armalite plastic and the lower 1.5m of mesh guarding. However, it has been found that 2m mesh provides satisfactory, inexpensive protection and that the 1.5m mesh with 0.5m armalite plastic is unnecessarily 'gilding the lily'.

The major robot installations

Derby Hauni Blohm line
This is a development of the Derby '7 cell line' described later. It is unguarded, but this has been achieved by having all of the robot operations localised within the machine guards.

Robot cut-off cells in foundries at Bristol and Derby
The robots are totally enclosed in sound and dustproof booths, and loading is via a rotating table. Access to the booths is interlocked.

Robot development areas at Derby and Bristol
Each of these development areas has 'robot pens' with 2m mesh fences. Access is interlocked to the robot controller.

Wheel and disc machining FMS at Derby
This flexible manufacturing system is known as AIMS (Automated Integrated Manufacturing System). AIMS will have AGVs and overhead loading systems for large machines. The AGVs will operate in an open shop, with proximity and personnel trigger guards which will halt the AGV if it comes into contact with an obstruction.

The overhead loader will be semi-automatic in that traverses will be controlled by operators in the manner of an overhead crane. Interlocks will prevent out of sequence operations being attempted.

Bristol 360° ECM line
This is a robot loaded 360° ECM process. It is fully guarded with 2m mesh fence and interlocked. All chemical handling is via an enclosed pump room to bulk process/storage facilities.

Fig. 1   The complete Derby '7 cell line'

Other developments
Other recent robot installations are a robot loaded broach line for compressor
blades and a radial grinding cell for grinding nozzle guide vanes.

Derby '7 cell line'

Although consisting of only five robot cells, it was originally planned to be a
seven cell line, and has been known as that ever since. Each of the five cells
consists of a pair of creep feed grinding machines, loaded by a mechanical
handling unit (pick and place robot). The components (turbine blades) are fed
from a central conveyor. The whole installation is guarded by a 2m mesh fence.
Access to each cell is controlled by interlocks which disable the robot if the

Fig. 2   Robot cell with an ASEA MHU

door is open. The system can only be overridden by authorised setters for maintenance and correction purposes.

Fig. 1 shows the complete '7 cell line' with its central conveyor. In particular, it will be seen how the machines may themselves become part of the perimeter guarding and also guard the central conveyor.

Fig. 2 shows a robot cell with an ASEA mechanised handling unit. The perimeter guarding is partly by open mesh and partly by the machines themselves. The conveyors enter the cell through a window which prevents human access. The door interlock may be seen at the top of the door on the far side of the figure. The machines themselves are accessible for routine maintenance without having to physically enter the cell.

Fig. 3 illustrates the conveyor guarding viewed form inside the cell. This guard also covers an orientation device inside the cell.

Fig. 3   Conveyor guarding as viewed from inside the cell

Acknowledgements

The author has written this robot safety case study with considerable help in the form of information, photographs, etc., from the staff of Rolls-Royce Ltd and a willingness to answer probing questions. In particular: Mr A. Jackson, General Facility Manager, WC-O, Derby, and Mr D.L. Carter, Manufacturing Engineering Manager – Derby Group of Factories, have provided the framework; and Mr A.J.S. Pratt, Acting Head of Advanced Manufacturing, and Mr E. Parrott, Manufacturing Engineering Manager – Automation and Technology, have provided the detailed information.

Associated Steels and Tools

B. Powis

Associated Steels and Tools Wolverhampton has a long history of supplying quality pressings and jig welded assemblies for UK car manufacturers. A large number of the assemblies require the jig welding of one or more pressed components to form the completed assembly. In the past, the operation was successfully carried out by skilled welders. However, a great proportion of the assemblies are produced in large quantities and on a regular production basis, thus requiring the installation of a robot welding unit. The robot would carry out the long and continuous welding operation, leaving the human welders to manufacture smaller quantity work.

System implementation and safety features

Much consideration was given to the choice of both the robot and it's associated equipment. The Unimation PUMA robot, supplied by Unimation (Europe) Ltd, Telford, coupled with an index turntable jig fixture manufactured by WRS Wolverhampton were selected. By choosing local firms it also enabled an easy programme of communication to be established.

The siting of the robot on the shop floor was also given careful consideration, with respect to the supply of components, the removal of finished components and to the supply of services to the enclosure. The existing welding shop was already well-positioned, with easy access from the press shop into the welding shop and easy access out of the welding shop into the painting, plating and despatch departments. The shop was also supplied with all the necessary services that would be required, i.e. correct electrical supply, compressed air, on-line gas supply for the welding operations.

One consideration was that with the robot situated in the required site, the existing building framework would form two sides of the required guarding enclosure. Services for the robot were also existing on these walls with both electrical and compressed air supplies being close at hand. The proposed site was serviced easily by the existing main gangways with adequate room for the supply and removal of components by the existing forklift truck operating throughout the factory. So with a little careful planning and forethought the robot installation had found its 'perfect home'.

Fig. 1 General layout

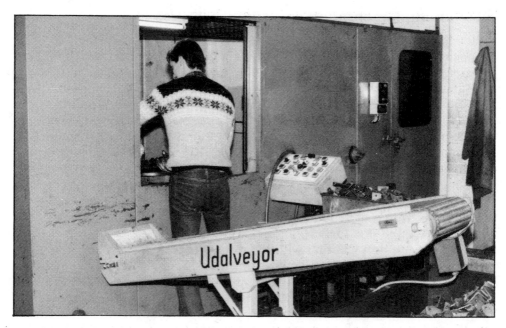

Fig. 2 Operator loading

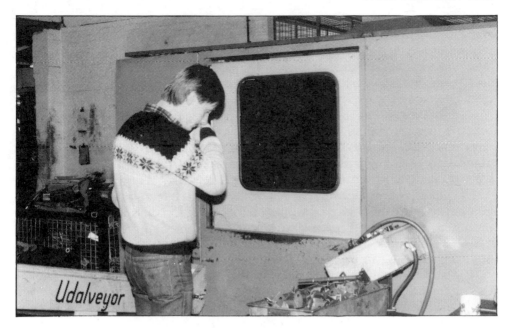

Fig. 3 Screen in fully closed position

Fig. 4 Side-door access into the robot enclosure

The framework of the building supplied two sides of the required robot enclosure, so the need for suitable guarding was considerably reduced. The remaining guarding system was undertaken by J.P. Udal Co., Wolverhampton, who are not only an associated company of Associated Steels and Tools, but also the largest mechanical guard manufacturer in the UK, with over 60 years of experience in the safe guarding of machinery.

Careful discussions took place with all personnel and a list of requirements for the correct and safe operation of the robot was drawn up. These requirements were also discussed with both the supplier of the robot and the associated equipment to ensure that the proposed safeguards and proposals were acceptable for correct machine interface.

The guard system that was chosen consisted of the remaining two open sides being fitted with 2m high partitioning panels. These panels were fixed to floor mounted uprights and formed a rigid enclosure around the robot. Careful consideration was given to the siting of the robot and the surrounding guarding, bearing in mind the possible reach of the robot arm when at it's extended reach – the company did not want the robot arm to damage itself or the enclosure when it was extended to full reach. This point was again considered when the 'services' were brought into the robot enclosure; all services were dropped down to floor level to avoid any flying leads or loops of cables pipes that may have become entangled in the robot's mechanisms.

As the application was for a welding fixture operation, the guard surround panels were fitted with viewing windows with suitable difussed glass to enable the viewing of the welding process from outside the enclosure.

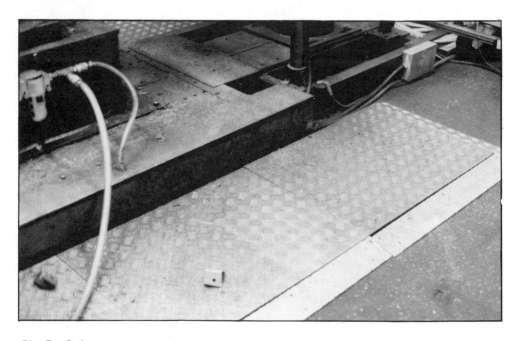

Fig. 5   Safety pressure mat

The load and unloading position of the operator was established and a horizontal sliding screen was detailed with suitable interlocking fitted. The screen was to be closed by an air cylinder, therefore a leading safety edge strip was fitted to the guard screen so that the guard would return to the open position if the screen met with an obstruction. The interlocking of the screen by suitable limit switches ensured that the screen had to be in the fully closed position before the machine could be operated.

Access into the robot enclosure was through a side access door interlocked with a Lowe and Fletcher key lock system. This involves the removal of one key from the control circuit, which removes power to the robot. The key is then used to gain access into the door interlock system; a second key has to be operated before the door can be opened. This operation locks the first key into the door interlock and prevents the key being removed. So we have a situation where once the door has been successfully opened power cannot be re-established to the robot. Only by closing the door and operating the keys in reverse can the master key be removed from the door interlock and power be restored to the robot arm.

The robot enclosure has now been safely guarded and a correctly interlocked loading point guard system and a correct access interlocking system has been allowed for. The safeguarding of the setter was then considered due to the need for the setter to have access to the robot arm and welding fixture during the 'teach mode'. Again a Lowe and Fletcher lock system was used to ensure that the master key had to be removed from the control circuit and inserted into the teach mode interlock. Once inserted and engaged the robot was put into teach mode and would only respond to the teach pendant. This again was suitably interlocked by a deadmans button and the whole system was in the power down condition, with only minimal power available for teaching.

Concern was still levelled at the possibility that a person could be effectively locked into the robot enclosure by a second person and become trapped inside the robot arms operating cycle. Therefore the floor of the enclosure was covered with a heavy duty safety pressure mat system, which again was suitably interlocked and would prevent the robot arm operating should the mat be operated (see page 205). The mat circuit consisted of an external reset feature which was operated by a security keyswitch fitted outside the robot enclosure. This meant that the pressure sensitive floor could only be reset by leaving the enclosure and then resetting the mat circuit. The control unit of the mat has a large colour coded illuminated display panel easily visible by all concerned.

This system would appear to cover the aspect of detecting a person inside the robot enclosure, but due to the application requiring the installation of a welding jig platform it was thought to be possible to have an individual balance on such a platform and therefore defeat the sensitive floor. To overcome this a system of infrared photoelectric sensors was fitted to cover the whole area of the robots operating arm. The sensors were fitted in a pattern of detection that would allow for the acceptable heights of the jigs and fixtures, and also the robot itself, but it would not accept the detection of any object that exceeded the profiled heights on the detected area.

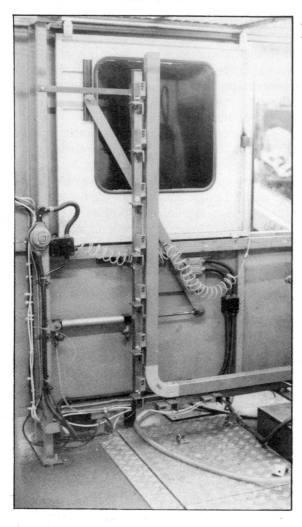

Fig. 6   System of photoelectric sensors within robot enclosure

This level of detection inside the enclosure is one of the most important parts of the safe guarding of the robot. Since the operator or setter is at the robot's mercy once he has entered the enclosure, an additional guard feature was added after the robot had been in service for a short time – a wire mesh on top of the enclosure to prevent the depositing of foreign objects over the side walls and onto the welding fixtures. These objects, such as paper cups, although not enough to stop the robot arm themselves, are detected by the infrared sensors and cause the unit to detect a fault inside the robot enclosure. Although not a great hazard it does go to show the level of detection that has been achieved inside the enclosure.

Once the robot had been installed and all the various safety devices fitted, a thorough and extensive training programme was implemented with the assistance of the robot manufacturer and various training programmes available through local services. All operators, toolsetters and maintenance staff to be involved with the robot participated. A more detailed training programme was carried out with the personnel selected for the actual programming of the robot and a full working knowledge of the safety features fitted was also implemented.

A careful monitoring system has been established with a daily record of safeguard checks and a correctly logged and duplicated system of computer programming, including all alterations and program changes. Refresher courses are planned on a regular basis to ensure that all personnel are kept up-to-date with the constant development of robots and their operating procedures.

Initially, the robot installation was condemned by some of the shop floor personnel. With all it's safeguards, interlocking and specialist training, they considered the whole thing to be far too complicated. But as time has passed by and the operating and power potential of the robot has been demonstrated, this attitude has now changed to one of respect.

Source of material

Robot safety and the law
First presented at the Robot Safety Seminar, 10–11 November 1982, Crick, Northants, UK. Organised by the University of Nottingham in conjunction with the Ford Motor Company.
Updated in February 1985. Reprinted courtesy of the author and HSE.

Safety standards in robotics
First presented at the Robot Safety Seminar (Univ. Nottingham/Ford Motor Co.)
Updated in March 1985. Reprinted courtesy of the author and MTIRA.

Systematic robot-related accidents and standardisation of safety measures
First presented at the 14th International Symposium on Industrial Robots, 2–4 October 1984, Gothenburg, Sweden.
Reprinted courtesy of the author and IFS (Conferences) Ltd.

Occupational safety and industrial robots
First presented at Ars Electronica '82, 28–30 September 1982, Linz, Austria.
Updated in September 1984. Reprinted courtesy of the author and IPA Stuttgart.

Robot accidents in Sweden
Report published by Arbetarskyddsstyrelsen, National Board of Occupational Safety and Health, Sweden, 1984.
Reprinted courtesy of the author and Arbetarskyddsstyrelsen.

People and robots: Their safety and reliability
First presented at the 7th British Robot Association Annual Conference, 14–16 May 1984, Cambridge, UK.
Reprinted courtesy of the authors and the British Robot Association.

Fault-tree analysis of hazards created by robots
First presented at the 13th International Symposium on Industrial Robots and Robots 7, 17–21 April 1983, Chicago, USA.
Reprinted courtesy of the authors and the Society of Manufacturing Engineers.

Safety system proposal for automated production
First published in abridged form in The FMS Magazine, January 1985.
Reprinted courtesy of the authors.

Design of safety systems for human protection

First presented at the 14th International Symposium on Industrial Robots, 2–4 October 1984, Gothenburg, Sweden.
Reprinted courtesy of the author and IFS (Conferences) Ltd.

Design for safeguarding

First presented at the RIA Robot Safety Seminar, 8–9 November 1984, Dearborn, USA.
Reprinted courtesy of the author and the Robotic Industries Association of America.

CAD – An aid to robot safety

First presented at the Robot Safety Seminar (Univ. Nottingham/Ford Motor Co.) Updated February 1985. Reprinted courtesy of the authors and the University of Nottingham.

Safety computer design and implementation

First presented at the Robots 8 Conference, 4–7 June 1984, Detroit, USA.
Reprinted courtesy of the authors and the National Bureau of Standards.

Increased hardware safety margin through software checking

First presented at the Robots 8 Conference, 4–7 June 1984, Detroit, USA.
Reprinted courtesy of the author, IBM, and the Society of Manufacturing Engineers.

Robot guarding – The neglected zones

Not previously published.

Safety interlock systems

Not previously published.

Photoelectric guarding

Internal report prepared by Manufacturing Safety Coordination, Ford of Europe, January 1985.
Reprinted courtesy of Ford of Europe.

Safety mats

First presented at the Robot Safety Seminar (Univ. Nottingham/Ford Motor Co.). Updated in January 1985. Reprinted courtesy of the author.

Perimeter guarding

Not previously published.

Safety sensor systems

First presented at the Robots VI Conference, 2–4 March 1982, Detroit, USA.
Reprinted courtesy of the author and the National Bureau of Standards.

A robot safety and collision avoidance controller

First presented at the Robots 8 Conference, 4–7 June 1984, Detroit, USA.
Reprinted courtesy of the authors and the Society of Manufacturing Engineers.

Case Studies

Erwin Sick
Not previously published.

British Leyland
First presented at the Robot Safety Seminar (Univ. Nottingham/Ford Motor Co.).
Reprinted courtesy of the author and BL Technology Ltd.

Ford Motor Company
First presented at the Robot Safety Seminar (Univ. Nottingham/Ford Motor Co.).
Reprinted courtesy of the author and Ford Motor Company.

Rolls-Royce Ltd.
Not previously published.

J.P. Udal Co.
Not previously published.

Authors' organisations and addresses

M.C. Bonney
Professor of Manufacturing
    Organisation
Department of Engineering
    Production
University of Technology
Loughborough
Leicestershire LE11 3TU
England

R.J. Barrett
Health and Safety Executive
McLaren Building
2 Masshouse Circus
Queensway
Birmingham B4 7NP
England

J.P. Bellino
Robotics and Vision Systems
    Department
General Electric Company
P.O. Box 17500
Orlando, FL 39860
USA

M. Bonfioli
Dipartimento di Meccanica
Politecnico di Milano
Piazza Leonardo da Vinci, 32
20133 Milano
Italy

Y.F. Yong
Managing Director
BYG Systems Ltd
Highfield Science Park
University Boulevard
Nottingham
England

P.E. Burton
ASEA Ltd
(formerly of BL Technology Ltd)
Industrial Robot Systems
25 Alston Drive
Bradwell Abbey
Milton Keynes MK13 9HA
England

J. Carlsson
Arbetarskyddsstyrelsen
National Board of Occupational
    Safety and Health
171 84 Solna
Sweden

L.R. Carrico
International Flexible Automation
    Center (INFAC)
210 Century Building
36 S. Pennsylvania Street
Indianapolis, IN 46204
USA

B.A. Cook
IBM Corporation
P.O. Box 1328
Boca Raton, FL 33432
USA

S. Derby
Rensselaer Polytechnic Institute
Troy, NY
USA

H.J. Duffy
Erwin Sick
Optic-Electronic Ltd
Lyon Way – Hatfield Road
St. Albans
Hertfordshire AL4 0LG
England

M.E.K. Graham
Herga Electric Ltd
Northern Way
Bury St Edmunds
Suffolk IP32 6NN
England

R. Jones
Centre for Robotics and
    Automated Systems
Imperial College of Science and
    Technology
Exhibition Road
London SW7 2BX
England

R.D. Kilmer
Industrial Systems Division
US Department of Commerce
National Bureau of Standards
Washington, DC 20234
USA

M. Linger
IVF (The Swedish Institute of
    Production Engineering Research)
Molndalsvagen 85
S-41285 Gothenburg
Sweden

Manufacturing Safety Coordination
Ford of Europe
Dagenham Training Centre
Dagenham
Essex RM9 6SA
England

P. Nicolaisen
Fraunhofer-Institut fur Produktions-
    technik und Automatisierung (IPA)
Nobelstrasse 12
D-7000 Stuttgart 80
West Germany

N. Percival
The Machine Tool Industry
    Research Association
Hulley Road
Macclesfield
Cheshire SK10 2NE
England

B. Powis
J.P. Udal Co.
Union Mill Street
Horseley Field
Wolverhampton WV1 3ED
England

N. Sugimoto
Research Institute of
    Industrial Safety
Ministry of Labour
5-35-1 Shiba
Minato-ku
Tokyo
Japan

N.K. Taylor
Department of Production
    Engineering and Production
    Management
University of Nottingham
Nottingham NG7 2RD
England

C. Thompson
Marconi Underwater Systems
(formerly of Ford Motor Company)
Bells Hill
Stoke Poges
Buckinghamshire
England

A.F.J. Yule
Safety Engineering
Ford Motor Company
Halewood Operations
Liverpool L24 9LE
England